MW01632696

Smart Cities as Democratic Ecologies

Smart Cities as Democratic Ecologies

Edited by

Daniel Araya
University of Illinois at Urbana-Champaign, USA

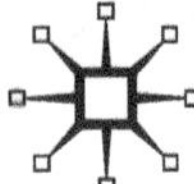

Selection and editorial matter © Daniel Araya 2015
Individual chapters © Respective authors 2015
Foreword © Nigel Jacob 2015
Foreword © Tim Campbell 2015

All rights reserved. No reproduction, copy or transmission of this publication may be made without written permission.

No portion of this publication may be reproduced, copied or transmitted save with written permission or in accordance with the provisions of the Copyright, Designs and Patents Act 1988, or under the terms of any licence permitting limited copying issued by the Copyright Licensing Agency, Saffron House, 6-10 Kirby Street, London EC1N 8TS.

Any person who does any unauthorized act in relation to this publication may be liable to criminal prosecution and civil claims for damages.

The authors have asserted their rights to be identified as the authors of this work in accordance with the Copyright, Designs and Patents Act 1988.

First published 2015 by
PALGRAVE MACMILLAN

Palgrave Macmillan in the UK is an imprint of Macmillan Publishers Limited, registered in England, company number 785998, of Houndmills, Basingstoke, Hampshire RG21 6XS.

Palgrave Macmillan in the US is a division of St Martin's Press LLC, 175 Fifth Avenue, New York, NY 10010.

Palgrave Macmillan is the global academic imprint of the above companies and has companies and representatives throughout the world.

Palgrave® and Macmillan® are registered trademarks in the United States, the United Kingdom, Europe and other countries.

ISBN: 978–1–137–37719–7

This book is printed on paper suitable for recycling and made from fully managed and sustained forest sources. Logging, pulping and manufacturing processes are expected to conform to the environmental regulations of the country of origin.

A catalogue record for this book is available from the British Library.

A catalog record for this book is available from the Library of Congress.

Contents

List of Figures

List of Tables

Foreword
The Evolution of Smart Cities

The notion of the Smart City has evolved. No longer simply a catchphrase for "faster, better, cheaper," smart cities are entering a new frontier in their evolution. Traditionally associated with regional competitiveness in the global economy, smart cities have now become embedded in a broad range of discussions on policy making and urban planning. As the thinking on this topic has deepened, so has our understanding of the ways in which people might live and grow *within* the Smart City.

As notions of smart cities have evolved, so has the language that is used to describe their potential. In fact, a new and augmented discourse has become prevalent in articulating a wide array of overlapping hopes and fears about the Smart City. New terms have arisen to express the social aspirations and social contradictions about smart cities in the 21st century. Increasingly, more sophisticated approaches to thinking and discussing the ways in which technology is being used to (re)make cities has emerged as a foundation to an evolving conversation on urban planning.

Exploring everything from social technologies supporting next-generation political institutions to new models of technology-mediated urban farming, smart cities have entered a new era. Notions of the Smart City have expanded to include cutting-edge public policies on urban economic clusters and new industries in the context of future urban design.

But this is only part of the story.

Human(e) Cities

The reality is that the ideal of smart cities has become quite fashionable in recent years. Against a background of strategic planning and policy modeling, we now see new frameworks that seek to leverage information and communication technology (ICT) and social capital in the growth of our living environments. Increasingly, smart cities are linked with concerns about environmental sustainability, human capital development, and urban governance.

But in truth, cities are a process. In fact, the Smart City is about more than technology. It is about balancing resilient communities with competitive industry in the process of making better-informed decisions. More than this, it is about moving decision making into the hands of residents so that they might contribute and apply their understanding to creating cities that are humane and just. Not just "competitive cities" or "creative cities," but cites that are livable, engaging, equitable, and fun.

All technologies have a system of values built into their very architecture. Top-down, hierarchical architectures keep power centralized and reduce consumers to end users at the edges. Beyond top-down cities, what we need today are peer-to-peer architectures that encourage more egalitarian approaches to power distribution. In fact, it is the *values* of Smart City technologies that must be a significant part of the conversation. If we wish to develop truly humane cities, we must encourage systems and technologies that give voice to the peoples and communities who make and remake the city every day.

Cities, like living organisms, evolve over time. They grow in response to how people build and use them. But not all changes to cities result in stable use patterns. Some city configurations provoke medical epidemics or social unrest, alongside elevated crime and social instability. The truth is that wise decision making can only arise from the deep insights found within the interaction of communities. Taking a wiser approach to decision making means investing in the proper resources to support social and technological systems that learn over time and place human scale values at the center. For this wisdom to have any material impact, it must be supported by real-world interventions – that is, policies and infrastructure that might enable dynamic communities.

Expanding the Concept of Smart Cities

In the current milieu, the concept of the Smart City evokes a wide range of images with regard to how technology may be used to improve or change the built environment. This kind of thinking often leads to conceiving cities as a system of systems. While this approach lends itself to many interesting and possibly useful abstractions, it misses something as well. It misses the critical importance of the people who live and shape cities. Building on the affordances of sensor technologies, data analysis, and urban design, we now have the potential to leverage newer and richer forms of democratic well-being, not just cities as engines of innovation but as ecologies of democratic collaboration.

Together, the authors in this anthology explore the cutting edge of the Smart City. As they remind us, the availability and quality of ICT infrastructure is not the only definition of the Smart City. Taking a special look at the role of human beings and human institutions in developing smart cities, this edited collection expands the concept of the Smart City to address the critical importance of democratic values in the continued evolution of smarter cities. In this, they are doing the fields of public policy and urban planning a great service. By expanding – perhaps even moving beyond – the concept of smart cities, the authors enable us to take a more critical view of a variety of possible futures for cities and city living.

Nigel Jacob
Co-Chair,
Mayor's Office of New Urban Mechanics
Boston, MA

Foreword
Toward Smart Cities as Democratic Ecologies

The starting point for the rich variety of perspectives in this volume is an acknowledgment of the rapidly changing terms of the social contract between citizens and the state in a smart urban world. The novel concept of democratic ecologies is the conceptual umbrella deployed to cover the wide-ranging topics included here. This is a welcome effort, addressing many interesting questions and raising many more – institutional, technological, political, and social – which face nations and cities in the coming decades.

At the risk of oversimplification, the chapters in this volume can be clustered in three broad areas. One small group of chapters takes an empirical stance, aiming to characterize the structural dimensions and operational features of smart cities by viewing them through distinctive prisms. One has sought to understand the extent to which the underlying features of green and resilient cities share commonalities with cities that are smart. Another gauges the power of peer-to-peer interaction. A third explores the potential and limitations of interactions on the urban edge. Each of these views adds new insight into the many shades of smartness.

A second group of authors takes a normative perspective. This group explores sponsorship of cities and the rules of the road by which citizen interactions should be conducted when using the communications and sensing technologies that are quickly permeating cities. Analysts cover questions of the role of government in building smart cities, the rise of surveillance, and the challenge of educating citizens to the new environment of smartness. Others in this group have explored, respectively, the impact of creative zoning and the opportunities and limitations of changing mobility. These chapters have begun to unpack the many facets of public policy that need to be explored as nations and cities undergo the next wave of technological transformation.

A third group blends normative and empirical approaches, but taking the perspective of citizens and users, rather than policy makers in smart spaces. These chapters combine speculative research and empirical analysis to understand the emerging qualities of smart cities, such as when their citizens and institutions are "always on" and completely interconnected. Young adult fiction and social media are tapped as sources of speculative analysis to probe the various kinds of narratives on the one hand, and on the other hand, looming concerns about privacy issues. One chapter explores how citizen input can be mobilized as source to evaluate emerging urban forms.

Each of these areas – what cities are, how they are to be formed, and how citizens operate in them – approximates the epistemological trek we must undertake in order to manage the new wrinkles, both positive and negative, as the domain of democracy widens in smart cities. Virtually all of the authors are clear about one thing: smart cities are not merely the aggregation of sensors, boxes with blinky lights and fiber-optic cable. The very premise of the book, and the many directions its contributors have indicated, reminds us that people in cities are the beginning and the end of the smart urban debate.

And yet the virtually complete penetration of handheld devices around the globe indicates the important extent to which smart cities are already here. The nearly 7 billion mobile platforms, whether or not they are connected to the Internet, constitute arguably the most important proto-tissue of smartness in cities. The next several decades will see remaining gaps close quickly in both coverage and capabilities of handheld and wearable devices.

The thick web work of exchange made possible by mobile devices alone constitutes the "wiring" for smart cities, without even counting the many billions of addressable objects already on the Internet and the tens of billions more that will be added in the coming decades. Cisco reminds us that the so-called Internet of everything is being adopted five times faster than was the case with the adoption of computers.

What is more, the youth bulge – the ripple of 15- to 25-year-olds that have emerged in the population pyramid as a result of public health successes – are already digitally literate. Their generation and the next one they spawn will live in an environment that generates more information every day than has been accumulated in all of history.

This leads us back to another aspect of democratic ecologies – participation in the city. The concomitant to an environment rich in information flows is complexity of participation in many new forms. Already, we are seeing new ways of taking part in the exchange of information, experimentation in new social and political norms, and innovation in the ways cities work.

Take the example of the camera and other features of low-cost mobile phones. Many are rightly concerned by official surveillance. Mobile technologies make this a two-way street. The term "sousveillance" was coined (by Steve Mann) to capture both citizen-initiated reporting of wrongdoing, as well as the idea of watching from below. Examples include the real-time reporting on the conduct of elections in Ghana and Sierra Leon.

There is no more compelling testimony to this short message service (SMS) transformation than the Arab Spring, particularly in Tunisia. Although set in a seething tinderbox of political suppression, and geared to national, not local regime, the power of the narrative offers a glimpse of future political activism on all levels.

At a more aggregated level, cities themselves are taking part in learning how to become smart. Thousands of cities are on the prowl around the globe to find and harvest ideas from other cities. The smartest among them create forms of social capital – what I call clouds of trust (networks of confidence) – to organize their learning, that is, to find new technologies and practices from peer cities, and to vet and adapt new ideas for use back home.

Citizens taking part in the formation of smart cities is one of the most exciting, if underaddressed, features of a wired urban world. That's why this volume is important. It begins to probe the manifold modalities of participation – individual (as persons), virtual (as intermediated by technology), and collective (as communities). These ideas are either manifest in or lurk behind much of the discussion in this volume. And there is so much still to explore.

Tim Campbell
Woodrow Wilson Center for Scholars
Washington, DC

Notes on Contributors

Barbara Adkins is a sociologist whose research focuses on the role of design in social and cultural life. She is an associate professor in the School of Design in the Creative Industries Faculty at the Queensland University of Technology (QUT) in Brisbane, and was formerly education manager and researcher at the Australasian Cooperative Research Centre for Interaction Design. She plays a leading role in the Design for Health and Social Inclusion research group at QUT. Her recent research has focused on the role of digital technologies and the designed environment more generally, in enhancing participation for people with disabilities.

Daniel Araya is a researcher and advisor to government with a special interest in education, technological innovation and public policy. His newest books include: *Augmented Intelligence* (2016), *Evolution of the Liberal Arts in the Global Age (2016), Rethinking U.S. Education Policy* (2015), and *Higher Education in the Global Age* (2013). He holds a doctorate from the University of Illinois at Urbana-Champaign and is an alumnus of Singularity University's graduate program at the NASA Research Park in Silicon Valley.

Hassan Arif is completing his PhD at the University of New Brunswick specializing in urban sociology and local government. He has worked as a research associate at the Urban and Community Studies Institute at the University of New Brunswick, and he has written as a columnist for the *Telegraph Journal*. His columns have appeared in the *Huffington Post* and *Sustainable Cities Collective*.

Michel Bauwens is the founder of the Foundation for Peer-to-Peer Alternatives and works in collaboration with various researchers in the exploration of peer production, governance, and property. He has coproduced the three-hour TV documentary *Technocalyps* with Frank Theys, and coedited the two-volume book on anthropology of digital society with Salvino Salvaggio. He is currently Primavera Research Fellow at the University of Amsterdam and external expert at the

Pontifical Academy of Social Sciences (2008, 2012). Michel is a member of the Board of the Union of International Associations (Brussels), and adviser to *Shareable* magazine (San Francisco) and Zumbara Time Bank (Istanbul). He functions as the chair of the Technology/ICT working group, Hangwa Forum (Beijing, Sichuan), to develop economic policies for long-term resilience, including through distributed manufacturing. He writes editorials for Al Jazeera English and is listed at #82, on the Post-Carbon Institute (En)Rich list.

Alice Birolo is a Master of Architecture candidate at Politecnico di Torino, Faculty of Architecture Construction and City (Turin, Italy). Currently, she is a visiting student at the MIT Senseable City Lab, where she is preparing her master thesis. Alice's research focuses on the concept of Smart City, and in particular on the development of smart islands in the city of today. More precisely, Alice's master thesis investigates a modular smart element for cities in order to respond to citizens' needs while being integrated into the urban architecture. Alice holds a Bachelor of Architecture degree from Politecnico di Torino. Her final dissertation was focused on the European Green Capitals Award with a deep comparison between the winner cities and Turin, one of the 2014 candidate cities.

Marianella Chamorro-Koc is a senior lecturer and the Subject Area Coordinator of Industrial Design at Queensland University of Technology (QUT), Brisbane, Australia. Her research work is within the areas of design, human factors/ergonomics, product usability, experiential knowledge, and context of use. Dr. Chamorro-Koc's research work aims to identify the experiential knowledge embedded in people's activities and interactions, and the contextual aspects shaping them. Her research interest focuses on understanding the ways in which people's practices of everyday life is influenced by their interactions with products, technologies, and systems. She is a researcher within the People and Systems Lab at QUT.

Matthew Claudel has a background in writing and studied architecture at Yale University, where he received the 2013 Sudler Prize – the highest award for creative arts. He now works in design, writing, education, and curation, with architecture projects in Tokyo and St. Kitts, and furniture in Denmark and Japan. He has written and coauthored several books, articles, criticism, and journalism; cotaught at MIT; lectured at

Harvard Business School; and presented widely as a critic, speaker, and artist-in-residence. He is an active protagonist of Hans Ulrich Obrist's 89plus, and collaborates on a broad array of projects. Matthew is currently a researcher at MIT's Senseable City Lab.

Isabel Cole, MIS, is a librarian and document specialist who has worked at the Internet Public Library and the Indiana State Library, as well as the Software Patent Institute and Thomas P. Miller & Associates. Her master's degree concentration was Library Science, which included study of referencing and cataloging. Her Master of Information Science degree is from the University of Michigan.

Roland Cole earned his BA in economics, master's and PhD in public policy, and a JD from Harvard University. Dr. Cole assists in developing SIPR programs and projects related to technology policy, identifying ways in which technology policy impacts existing SIPR research projects. Since 1992, Cole has served as executive director of the Software Patent Institute (SPI), an Indianapolis-based nonprofit that operates an online database of key documents and offers courses to clients such as the U.S. Patent and Trademark Office. He has also served as an attorney at Shughart Thomson & Kilroy in Kansas City, Missouri, and at Barnes & Thornburg in Indianapolis. He is a cofounder of the international Association of Personal Computer User Groups (APCUG). In addition, Cole has coauthored a number of books and has taught at the University of Michigan, Indiana University (Indianapolis), and the University of Washington.

Bill Cope is a research professor in the Department of Educational Policy Studies, University of Illinois, Urbana-Champaign, and an adjunct professor in the Globalism Institute at RMIT University, Melbourne. He is also a director of Common Ground Publishing, developing and applying new publishing technologies. He is a former first assistant secretary in the Department of the Prime Minister and Cabinet and Director of the Office of Multicultural Affairs in Australia. His research interests include theories and practices of pedagogy, cultural and linguistic diversity, and new technologies of representation and communication. He is working on a project investigating the next generation of "semantic web" technologies.

Gerry Derksen is an associate professor at Winthrop University in South Carolina and coordinator of the Interactive Media program in the Department of Digital Information Design. He holds degrees from the University of Manitoba (BID) in the Department of Architecture and from the University of Alberta (MDes) in the Department of Fine Art and Design. His areas of research include usability and heuristic evaluation, experience design, and natural language processing. His projects focus on conversation modeling, sensor technology in human computer interaction, and computational narrative generation. Most notable papers stem from his work on reenactments of crime scenes and their prejudicial impact on a jury: "Film as Surrogate, Effect of Visual Elements on the Perception of Jurors in Computer Re-enactments of Crime Scenes," and "Jury Perceptions of Animation and the Uncanny Valley."

Mary Kalantzis is Dean of the College of Education at the University of Illinois, Urbana-Champaign. Before this, she was Dean of the Faculty of Education, Language and Community Services at RMIT University, Melbourne, Australia, and President of the Australian Council of Deans of Education. She has been a board member of Teaching Australia: The National Institute for Quality Teaching and School Leadership, a commissioner of the Australian Human Rights and Equal Opportunity Commission, chair of the Queensland Ethnic Affairs Ministerial Advisory Committee, vice president of the National Languages and Literacy Institute of Australia, and a member of the Australia Council's Community Cultural Development Board.

Ayesha Khanna is an education, technology and urbanization expert with over 15 years of experience at the leading edge of innovation from smart city development to brand strategy. She is the co-founder and CEO of The Keys Academy, an innovative enrichment hub whose unique model of "externships" with leading companies provides secondary school students the opportunity to apply their skills to critical 21st century industries. In 2014, Ayesha also served on the Singapore Ministry of Education's ASPIRE Steering Committee on higher education reform and applied learning. She is founder and chairman of Applied Skills, a provider of digital platforms and e-learning content for corporations. She is also co-founder and chairman of Factotum, a boutique content marketing agency that creates thought leadership branding for companies and governments.

Parag Khanna is a leading global strategist, world traveler, and best-selling author. He is a managing partner of Hybrid Reality, senior fellow at the New America Foundation, senior fellow at the Singapore

Institute of International Affairs, and adjunct professor in the Lee Kuan Yew School of Public Policy at the National University of Singapore. He is coauthor of *Hybrid Reality: Thriving in the Emerging Human-Technology Civilization* (2012), and author of *The Second World: Empires and Influence in the New Global Order* (2008) and *How to Run the World: Charting a Course to the Next Renaissance* (2011). In 2008, Parag was named one of *Esquire*'s "75 Most Influential People of the 21st Century," and featured in *WIRED* magazine's "Smart List." Parag is regularly featured in media around the world such as the *New York Times, TIME, Financial Times,* and *Wall Street Journal,* and appears regularly on CNN, BBC, Al Jazeera, PBS, and NPR. He holds a PhD from the London School of Economics, and bachelor's and master's degrees from the School of Foreign Service at Georgetown University. Born in India, Parag grew up in the United Arab Emirates, New York, and Germany. He has traveled to more than 100 countries and is a Young Global Leader of the World Economic Forum and a Fellow of the Royal Geographical Society.

Tony Kim works as a trusted adviser to senior executives of large public sector organizations and enterprises where transformation is imminent or there is the potential for groundbreaking innovation projects. In the Cisco Consulting Services, Global Public Sector practice, Tony's focus is on Digital Business Platforms, Big Data/Analytics, Work and Learning Future, Smart + Connected Communities, and Connected Government projects in Asia-Pacific countries. Based in Seoul, Tony works with both government and large enterprises to develop organizational transformation and sustainable growth strategies. He works as trusted adviser to senior government leaders and management executives to rethink policy and growth strategy, and to transform value chain and business models through technology-enabled innovations. Tony provides thought leadership and reference practices for shaping the policy on government transformation and launching national initiatives – Gov 3.0 and Creative Economy. He provides consulting services to build the national smart work strategy and to co-create smart work readiness assessment program. He also focuses on big data/analytics projects while providing consulting services to innovate public services and enterprise business models.

Vasilis Kostakis is a senior research fellow at the Ragnar Nurkse School of Innovation and Governance, Tallinn University of Technology, Estonia. He is also founder of the interdisciplinary research hub P2P Lab and longtime collaborator of the P2P Foundation.

Kerry Mallan is professor and director of the Children and Youth Research Centre at Queensland University of Technology, Australia. She has published extensively on children's texts and cultures. Her most recent book is *Secrets, Lies and Children's Literature* (2013) published by Palgrave Macmillan.

Anijo Punnen Mathew is an associate professor at IIT Institute of Design. His research looks at evaluating new semantic appropriations of place as enabled by technology and media convergence. He also works with companies to adapt and change to the information economy through the use of design methodology and theory. Anijo holds a PhD in computing from the Open University in the United Kingdom, a Master of Design Studies (MDesS) from Harvard University's Graduate School of Design, and a professional Bachelor of Architecture (BArch) from Birla Institute of Technology, Mesra, Ranchi (India). Anijo has consulted and conducted research for many cities and city-based organisations such as City of Chicago, Hong Kong Design Centre, Singapore Urban Redevelopment Authority/ Singapore River One, Chicago Loop Alliance, and the Chicago Architecture Foundation. In 2007 the Architectural Research Centers Consortium (ARCC) selected Anijo as their New Researcher of the Year. He currently serves on an executive committee for The Chicago Convention and Tourism Bureau (Choose Chicago), and is on the advisory boards of the Chicago Architecture Foundation (CAF), and the Chicago Loop Alliance (CLA).

Piotr Michura is a lecturer at the Department of Design Fundamentals, Faculty of Industrial Design, Academy of Fine Arts in Krakow, Poland. He is a graduate of the Academy of Fine Arts in Krakow, Poland (MFA in Visual Communication Design 2001, PhD 2012) and the University of Alberta, Canada (MDes in Visual Communication Design 2008). His PhD research was on experimental visualizations for interaction with electronic documents for humanities researchers. His current research interests focus on conversation modeling theory and practice in interaction design fostering agency and engagement of stakeholders and generative potential of contextual factors.

Vasilis Niaros is an engineer/urbanist and a PhD student at the Ragnar Nurkse School of Innovation and Governance, Tallinn University of Technology. He is also a research fellow at the P2P Lab.

Carlo Ratti is director at MIT Senseable City Lab and Partner at Carlo Ratti Associati. An architect and engineer by training, Carlo Ratti practices in

Italy and teaches at MIT, where he directs the Senseable City Lab. Ratti has coauthored more than 250 publications and holds several patents. His work has been exhibited in several venues worldwide, including the Venice Biennale, MoMA in New York City, and MAXXI in Rome. At the 2008 World Expo, his "Digital Water Pavilion" was hailed by *Time* magazine as one of the "Best Inventions of the Year." He has been included in *Blueprint* magazine's "25 People who Will Change the World of Design" and in *Wired* magazine's "Smart List 2012: 50 People who Will Change the World." He is curator for the "Future Food District" at Expo Milano 2015.

Stan Ruecker is an associate professor at the IIT Institute of Design, with current research interests in the areas of humanities visualization, the future of reading, and information design. His work to date has focused on developing prototypes to support the hermeneutic or interpretive process, and he has published extensively on information design, experimental interface design, and interdisciplinary research project management. His book *Visual Interface Design for Digital Cultural Heritage,* coauthored by Milena Radzikowska and Stéfan Sinclair, was released in 2011. He holds an interdisciplinary PhD in Humanities Computing from the University of Alberta, an MDes from the same university, an MA in English Literature from the University of Toronto, and advanced undergraduate degrees in English Literature and Computer Science from the University of Regina. He is an adjunct professor at the University of Alberta's Humanities Computing Program, and in the University of Victoria's Faculty of Humanities.

Michelle Selinger set up as an independent consultant after leaving Cisco Consulting Services in December 2013. She offers strategic advice to support education transformation and redesign across all areas of education, online and in situ, and she undertakes reviews and evaluations of education programmes. Michelle has an academic background in education, particularly in technology enhanced learning, and has extensive experience helping governments and educational institutions around the world in both developed and emerging economies to create and evaluate strategies for education reform. Her experience includes acting as an evaluator, project reviewer and an expert adviser to a number of international agencies including the European Commission, UNESCO, the World Economic Forum's Global Education Initiative, the South African government on national e-skills development. Michelle also served as a member of the Australian Federal Minister of Education's Digital Education Advisory Group.

Olga Smirnova is an assistant professor at the MPA program, the Department of Political Science, East Carolina University. Her research interests include transportation, green transportation innovations, institutional stability, economic development, social networks, performance measurement, and visualizations of complex systems. She has published in *Public Administration Review, Journal of Public Transportation, Southeastern Geographer, Municipal Finance Journal,* and *North Carolina Geographer.*

Lisa Stafford is an academic and researcher in the School of Design, Design for Health and Social Inclusion research group at Queensland University of Technology, Australia. Her research interests are universal design and person–environment studies, with a particular focus in children's participation, mobility, and inequalities.

Kevin Stolarick is research director at the Martin Prosperity Institute at the Rotman School of Management, University of Toronto. He has held faculty positions at the College of Humanities and Social Sciences and the H. John Heinz III School of Public Policy and Management, Carnegie Mellon University, Pittsburgh, Pennsylvania, and for over a decade worked with technology in the insurance industry as a manager of strategic projects. He has also worked as a researcher at the Heinz School's Software Center, one of the Sloan Industry research centers, and as a "special sworn researcher" at the Census Data Research Center and a "deemed employee" at Statistics Canada.

Tarun Wadwa is research fellow with Singularity University, where he studies digital identification systems and their impact on public distribution, banking, and governance. He is also senior research associate with the Think India Foundation, where he analyzes the issues and challenges that India is facing due to urbanization and a researcher with the Hybrid Reality Institute, a research and advisory think tank focused on the intersection of technology trends and geopolitics. He is also on the Transparency and Privacy, and Cybercrime/Malware Advisory Boards for the Lifeboat Foundation. He is a contributor to *Forbes, Singularity Hub,* and *The Huffington Post*. His work has been published in several major media outlets and he has given talks at major universities, conferences, and related events.

Introduction

Daniel Araya and Hassan Arif

There are now over 400 cities in the world with a population of more than 1 million residents and over 20 cities with a population of more than 10 million residents. This remarkable urban growth has created vast policy and planning challenges related to infrastructure, governance, and environmental sustainability. Alongside mounting challenges related to power grids, transportation networks, and water distribution systems, there is an acute need to begin to examine the potential of "smart cities" as sites for social and political transformation. Beyond the Fordist city and its structural dependency on encrusted bureaucratic systems, the smart city represents an idealized attempt to reconstruct cities in the wake of techno-economic change (Perez, 2002). Indeed, the emergence of global city networks "as a transnational force beyond the top-down world of international negotiations or the bottom-up advocacy of civil society groups" (Acuto, 2013: 486) has begun to accelerate discussions on policy and planning vis-à-vis a wide array of social and political challenges.

Linking issues of open government and open data, for example, this collection explores the evolution of smart technologies in rethinking democratic institutions and practices (Benkler, 2006; Noveck, 2009). Drawing on the work of leading scholars, this edited volume focuses attention on changes in the relationship between citizenship and technological innovation in the context of "democratic ecologies." The term *democratic ecology* is used to connote the city as a living environment where communication and joint decision making are embedded within the design of cities.

Given the fact that an estimated 70 percent of the world's population will live in cities by the year 2050, it only makes sense that smart cities have become a key feature of the discourse on urban policy and planning (United Nations, 2014). One of the key challenges confronting the development of smart cities, however, is their *top-down* design. Rather than an abstract ideal promoted by policy makers or economists, smart cities might be better conceived of as living habitats capable of adapting to the emergent needs of citizens and residents.

Examining the growing potential for expanding democratic agency within smart cities, the authors in this collection explore the contours of a new era in urban planning. This includes new avenues for promoting government accountability, and new tools for enabling collaborative democracy (Goldstein and Dyson, 2013; Noveck, 2009). Much as Ayesha and Parag Khanna suggest (this volume), cities now need to enable government efficiency and public access to useful data. This includes public policies built on top of platforms linked to e-government services and government transparency. Indeed, as human habitation becomes increasingly urbanized, the world's major cities are becoming anchors for networks of "strategic governance" that are increasingly critical to resolving overlapping social, economic, and environmental challenges (Healey, 2002).

ICTs and democracy

The truth is there remains a paucity of literature on smart cities that explores issues associated with democratic agency and governance. Indeed, the very idea of smart cities has been widely criticized as little more than neoliberal corporatization in disguise (Greenfield, 2013). Hollands (2008), for example, critiques the neoliberal tendencies of smart city planning, particularly the empowerment of corporate and business actors at the expense of democratic governance. This is unfortunate because the world's cities are now the key drivers of policy and planning across a wide range of transnational issues (Acuto, 2013).

As many scholars and analysts agree, the corporatized design of smart cities puts an excessive weight on economic values as the sole driver of urban development (Townsend, 2013). Notwithstanding utopian visions of high-tech smart cities, the widespread use of ICTs as tools for surveillance suggests that smart technologies may well signal a new era in the design and evolution of autocratic power structures. Indeed, Foucault's (1975) notion of modern societies as "panopticons" suggests that smart technologies deployed in the design of smart cities must be evaluated in terms of how they enable (or curtail) human rights and freedoms. Indeed, what are the consequences of layering public spaces with computationally mediated technologies? And what is the potential of smart cities as platforms for bottom-up civic engagement in the context of next- generation communication, data sharing, and application development?

What is apparent is that the technologies that make possible the comprehensive accumulation of data offer both government and commercial industry an unprecedented capacity for social and political surveillance. This is not to say that technology determines the city but that it provides a platform for transforming civic practices. The danger is that the tendency of smart city initiatives to reinforce corporatized policy and planning will undermine the credibility of smart cities as spaces for citizen empowerment.

In considering cities as ecologies, for example, designers must reconcile multiple social needs as well as the allocation of sustainable resources. This includes the conscious design of smart technologies that offer urban residents greater agency in collaborative decision making. Greenfield (2006), for example, proposes the idea of linking ubiquitous computing to "open public objects" facilitating new forms of citizen informatics. Highlighting the capacity of ICTs to facilitate citizen empowerment, Tapscott and Williams (2006: 1) point to new technologies as tools for providing emergent models of social engagement "based on community, collaboration, and self-organiz[ation] rather than on hierarchy and control." As these thinkers observe, the potential of ICTs to move cities beyond the centralized planning of "the industrial age" and toward self-organizing systems of governance offers an unprecedented opportunity to rethink government (Tapscott and Williams, 2010).

While governments in the industrial age were based on the "monopoly of power, and structured around rigid hierarchies, today's governments need to distribute power broadly and leverage innovation, knowledge and value from the private sector and civil society" (Tapscott and Williams, 2010: 263–264). Indeed the development of ICTs for purposes of e-government has now become a kind of gold standard for promoting more open and efficient government. ICTs represent new sociotechnical tools that challenge the legitimacy of older models of social governance. Perhaps most important, the proprietary technologies underlying smart urbanization now offer corporations and business enterprises significant influence at the expense of democratic accountability.

Put simply, it is becoming increasingly important to consider new ways in which smart control systems can be subordinated to democratic institutions and practices. This reflects tensions in the use of ICTs vis-à-vis corporate planning models (Sassen, 1996) and the privatization of public governance across a wide swath of public services (Stoker, 1998). In response to this tension, a rich discourse on "Open Government" has emerged as a key feature of a new approach to rethinking civil society and the public sector itself (Goldstein and Dyson, 2013; Lathrop and Ruma, 2010). What this discourse suggests is that the potential of smart technologies to transform municipal governance, services, and land use – themes that are highlighted in the chapters of this book – points to a possible shift away from the top-down municipal bureaucracies. As Jane Jacobs has suggested, the potential for a new, more democratic era in municipal governance is critical to the future of cities. More than situated places to live, cities are becoming processes of social and economic transformation within a global network (Castells, 1996).

In critiquing smart cities, one might consider the work of Jacobs and other residents in New York City and Toronto, and their civic opposition to top-down planning models. As Jacobs demonstrated, top-down mega projects (freeways) can often undermine the vibrancy of the neighborhoods they are meant to serve. Rather than generating antagonism, smart cities could be designed to enable recursive learning and communication between citizens and their government. In Chicago, for example, EveryBlock (available online: http://chicago.everyblock.com) offers Chicago residents to track crime in different neighborhoods through a combination of Google online maps and the city's crime statistics. In Los Angeles, Neighborhood Knowledge California (http://nkca.ucla.edu) identifies neighborhoods at risk through factors such as tracking tax delinquency and fire violations, as well as providing data to the public enabling collaboration. Tools like these have the potential to democratize the urban policy-making process.

The truth is that bridging the interests of citizens and government requires a sufficient emphasis on democracy. Indeed, smart technologies could enable citizens to assess the effectiveness of their city's law enforcement policies, and measure the claims of politicians against realities on the ground. Rather than enhancing surveillance and top-down control, ICTs could be employed toward greater democratic accountability and citizen empowerment.

Whether ICTs promote or inhibit collaborative democracy depends upon the degree to which smart technologies enhance the political agency of citizens. The challenge today is to ensure that the potential of the smart city is not sacrificed to imperfections in design and implementation. Smart cities as code for policies focused on attracting investment at the expense of the public sector are simply no solution to widening social and economic disparities (Hassan, 2004). Unfortunately, the dominant role of private corporations in the application of proprietary technologies to the design and development of smart cities means that the potential for serious abuses of privacy, autonomy, and political agency expands considerably.

Notwithstanding the fact that debates on smart cities are largely fixed on technology, it remains the case that the underlying social and political potential of smart cities remains to be fully realized. As the authors in this collection suggest, the debate on smart cities should not be seen as a litmus test on the legitimacy of smart technologies for transforming urban environments. More than simply empowering citizens to provide greater input into highly centralized institutions, smart technologies could point the way to (re)designing governance structures and decentralizing decision making. Indeed, the erosion of public governance in the neoliberal era has motivated a host of social actors, including individuals, the volunteer sector, and nongovernmental organizations (NGOs), to assume new roles in establishing greater citizen and public input as well as connect people and organizations. These additional actors can and should challenge the momentum of neoliberal processes within smart cities.

With these cautions in mind, and remembering the different and contrasting ends that ICTs can be employed toward, it is nonetheless important to recognize the social and political benefits of smart cities. This includes new frontiers in open government and new means for lessening – if not entirely eliminating – top-down tendencies in policy development and implementation. These are some of the themes that animate the chapters of this book.

Overview of the book

Notwithstanding the growing problems associated with the top-down design of smart cities, this collection focuses on the emergent features of governance and policy making that might enhance democratic self-government. It begins with an examination of the new trends now shaping the design and discourse on citizenship and governance in smart cities. It then explores urban planning issues associated with the political and social evolution of smart cities. Last, it considers some of the features needed to improve upon contemporary urban design.

In Chapter 1, "Smart Cities and the Network Society: Toward Commons-Driven Government," Araya argues that the growing trend toward open government data suggests that a new era in commons-driven governance is on the horizon. What is now obvious, he suggests, is that the push for smart technologies is driving an expansion in surveillance systems (both public and private). Arguing for the need to consider new forms of democracy in the context of smart cities, he questions the ethics of closed institutions. More important, he links the notion of Open

Government to commons-driven governance and the empowerment of citizens within democratic ecologies.

In Chapter 2, "Government's Role in Growing a Smart City," Ratti, Claudel, and Birolo examine the consequences of integrating pervasive digital technologies with urban environments. As they point out, a broad spectrum of implementation models are emerging in different parts of the world. Nonetheless, questions remain regarding the role of government within smart city design and development. How can smart city funding be used most effectively, specifically to promote innovation?

In Chapter 3, "The Generative City," Ayesha and Parag Khanna examine the increasing importance of smart cities in the context of public policy. They argue that rapid urbanization, aging infrastructure, and scarce financial resources demonstrate the need for technological innovation. More than technology alone, however, smart cities should also consider investments in policy and governance.

In Chapter 4, "Urban Research Machines: Engaging the Modern Urban Citizen," Mathew argues that civic engagement has always played a key role in shaping a city's future. The challenge today, he suggests, is to reach out to the modern urban citizen through new models of interaction and communication. To this end, he examines Urban Research Machines that engage citizens with new tools of social participation and spectatorship. According to Mathew, these machines reimagine user engagement within the context of participatory environments that allow urban audiences to "play" with information and take collective action.

In Chapter 5, "Conversation and Narrative in the Smart City," Derksen, Michura, and Ruecker consider the implications of viewing the smart city as a reading environment. As various kinds of data are combined and gathered through digital technologies, they consider the potential for developing new and interesting uses of text within the city. The authors discuss two experimental platforms that could support enhanced decision making and democratic governance. The first platform is the Data Stories project, which sequences text from various dynamic sources through a thematic clustering algorithm (Latent Dirichlet Allocation). The second platform models linear discussions as 3-D objects via Technology Mediated Conversation Modeling (TMCM).

In Chapter 6, Stolarick and Smirnova ask the question, "Are Creative and Green Cities Also Smart and Clean?" Building on new research data, they explore the features and variables that distinguish the discourses on urbanization today. Their research identifies and compares creative, green, and smart cities, looking for correlations. As they conclude, cities that rank well in one category (e.g., green cities) seem to also rank well in other categories. Examining statistical outliers, they consider which cities are creative but not smart, and/or green but not creative. For cities that are simultaneously creative, green, and smart, they seek to understand the underlying relationships driving these results.

In Chapter 7, "Urban Reconfiguration after the Emergence of Peer-to-Peer Infrastructure: Four Future Scenarios with an Impact on Smart Cities," Kostakis, Bauwens, and Niaros consider conflicts emerging from the control and uses of peer-to-peer (P2P) networks as both an infrastructure and a new means of production. They introduce four future scenarios for economy and society (netarchical capitalism, distributed capitalism, resilience communities, and global commons). They explore the possible evolution of smart cities in each of these contexts. In their view, P2P-driven systems allow for alternatives to capital accumulation, while enabling knowledge and innovation diffusion through true citizen engagement.

In Chapter 8, "Smart Cities: Toward the Surveillance Society," Wadhwa weighs the consequences of smart technologies and their significance for citizens within democratic societies. With major opportunities for new innovations in urban governance, he considers the implications of collecting the personal information used in creating the smart city experience. As cities become "smarter," more and more personal data are collected through vast networks of sensors. Is the "smart city" simply the institutionalization of a vast surveillance society?

In Chapter 9, "Surviving the Electronic Panopticon: New Lessons in Democracy, Surveillance, and Community in Young Adult Fiction," Mallan considers the impact of digital technologies in shaping the social imagination. What are the implications of vast systems of data gathering underlying the design of smart cities? As modern societies become more mobile and networked, we see the increased surveillance, tracking, and spreading of dis/information. With the acceleration of new pervasive and immersive technologies, these questions have taken on new urgency and significance that go beyond an Orwellian Big Brother scenario. Looking closely at young adult fiction, Mallan extends Foucault's notion of the panopticon to take account of the challenges confronting networked democracies.

In Chapter 10, "Smart Cities Need Smart People," Kim and Selinger explain Cisco's strategic planning in the development of Songdo International Business District in South Korea. Building on a pervasive network infrastructure, the authors explain that digital networks in Songdo form a fourth utility that might support active and engaged citizens with robust access to information and services. As they explain, learners will develop the skills they need to succeed in the future, through a ubiquitous connectivity that extends beyond the walls of educational institutions.

In Chapter 11, "Zoning Experiments for Smart Cities," Roland Cole, Cole and Arif argue that smart cities need to focus on more "variable" density than they do currently. As they explain, the problem is simply that space costs and travel time severely undercuts the benefits of diverse interactions. In their view, cities need an experimental approach in urban planning to allow and encourage experiments in mixed use. Their chapter explores the experiments already taking place along these dimensions, with zoning laws and incentives encouraging experiments that mix public and private spaces, high and low density, and the "accidental mingling" that leads to smart collaboration, innovation, and improvements in quality of life.

In Chapter 12, "Designing New Mobilities for Accessible Cities: Scenarios for Seamless Journeys," Adkins, Chamorro-Koc, and Stafford consider issues around universal design. While cities now deliberately attempt to make resources accessible for people with disabilities, the realities of trying to access the journeys and connections considered integral to urban life continue to be frustrating and prohibitive for many. Applying the new mobilities paradigm, this chapter discusses design scenarios that consider the role of movement, time, and space in support of situated democratic environments.

In Chapter 13, "ExtraUrbia, or, the Reconfiguration of Spaces and Flows in a Time of Spatial-Financial Crisis," Cope and Kalantzis propose a repositioning of the urban in relation to five spaces they call extraurban. These are edge-urban, de-urban, micro-urban, greenfield, and off-the-grid. While the urban may be favored as a force for social transformation, they identify new continuities across "extraurbia" and argue that these new spatial flows are symptoms of the recent spatial-financial crisis.

Conclusion

Much as discourses on the knowledge economy favored an educated "creative class," so discussion on smart cities today tends to focus on the rising global influence of elite cities and their most affluent residents. In many ways, this edited volume is a response to this elite discourse, highlighting the need for changes in urban governance in support of the widest possible distribution of democratic practices. In this way, smart cities and the development of democratic ecologies might be seen as not merely opening government to citizen input but transforming the system of governance itself through the application of technology and the reform of democratic practice.

Challenging older assumptions undergirding urban design and planning, smart cities may well be facilitating new avenues for greater citizen collaboration, more open government, and better planning for social and environmental sustainability. At the same time, the problems associated with corporate hegemony over these technologies – and the systems of surveillance they make possible – remain to be overcome.

References

Acuto, M. (2013). City leadership in global governance. *Global Governance,* 19(3), pp. 481–498.

Benkler, Y. (2006). *The wealth of networks.* Princeton, NJ: Princeton University Press.

Castells, M. (1996). *The rise of the network society.* Oxford: Blackwell.

Foucault, M. (1975). *Discipline and punish: The birth of the prison.* New York: Random House.

Goldstein, B., and Dyson, L. (eds.). (2013). *Beyond transparency: Open data and the future of civic innovation.* San Francisco: Code for America.

Greenfield, A. (2006). *Everyware: The dawning age of ubiquitous computing.* Berkeley, CA: New Riders.

Greenfield, A. (2013). *Against the smart city.* Retrieved from: http://urbanomnibus.net/2013/10/against-the-smart-city/

Hassan, R. (2004). *Media, politics and the network society.* Maidenhead, UK: Open University Press.

Healey, P. (2002). On creating the "city" as a collective resource. *Urban Studies,* 39(10), pp. 1777–1792.

Hollands, R. (2008). Will the real smart city stand up? Creative, progressive, or just entrepreneurial? *City,* 12(3), pp. 303–320.

Lathrop, D., and Ruma, L. (Eds.). (2010). *Open government: Collaboration, transparency, and participation in practice.* Sebastopol, CA: O'Reilly Media.

Noveck, B. S. (2009). *Wikigovernment: How technology can make government better, democracy stronger, and citizens more powerful.* Washington, DC: Brookings Institution.

Perez, C. (2002). *Technological revolutions and financial capital: The dynamics of bubbles and golden ages.* Cheltenham, UK: Edward Elgar. Retrieved from: http://www.carlotaperez.org/pubs?s=tf&l=en&a=technologicalrevolutionsandfinancialcapital

Sassen, S. (1996). Whose city is it? Globalization and the formation of new claims. *Public Culture,* 8(2), pp. 205–223.

Stoker, G. (1998), Governance as theory: Five propositions. *International Social Science Journal,* 50: 17–28.

Tapscott, D., and Williams, A. (2006). *Wikinomics: How mass collaboration changes everything.* New York: Portfolio.

Tapscott, D., and Williams, A. (2010). *Macrowikinomics: Rebooting business and the world.* New York: Penguin.
Townsend, A. M. (2013). *Smart cities: Big data, civic hackers, and the quest for a new utopia.* New York: Norton.
Tranos, E., and Gertner, D. (2012). Smart networked cities? *Innovation – The European Journal of Social Science Research*, 25(2), pp. 175–190.
United Nations. (2014). *World urbanization prospects: The 2014 revision.* New York: Department of Economic and Social Affairs, United Nations.

1
Smart Cities and the Network Society: Toward Commons-Driven Governance

Daniel Araya

Since the launch of technology-driven infrastructure projects like IBM's Smart Planet and Cisco's Smart Communities, interest in smart city planning has grown substantially. Spanning a wide range of discussions on urbanization, the concept of the *smart city* overlaps a wide-ranging discussion on contemporary socioeconomic development. Despite its expanding influence, however, there is little consensus on the precise meaning of a smart city. While for some, the smart city refers to advances in sustainability and green technologies, for others, it denotes the deployment of information and communication technologies (ICTs) as next-generation infrastructure. One reason for the ambiguity is that the concept of the smart city means different things to different disciplines. Indeed, the push for smart cities has introduced a host of social policy concerns linked to neoliberal urban planning. Advancing on a corporate discourse that reimagines ICT platforms as cybernetic management systems, smart cities are now advertised as the *future of globalization*. Building on a critique of this discourse, this chapter focuses on a third strand in the discussion on smart cities. Linking rising demands for participatory democracy to ongoing discussions on smart cities, I explore the political ramifications of network technologies for reshaping democratic government.

Beyond smart cities

Given the unprecedented migration of much of the world's population into cities, it only stands to reason that cities have become central to public policy discussions across a host of social, economic, and political challenges. Global urbanization has swelled from 746 million in 1950 to 3.9 billion in 2014. By 2050, 66 percent of the world's population are expected to live in cities (United Nations, 2014) with most of this explosive growth occurring in developing countries.

Perhaps the central feature of the discourse on smart urbanization is the desirability of technologies to monitor and guide human behavior. Indeed, the "smartness" of the smart city lies in the circulation of data through vast webs of hardware and software. This includes feedback systems rooted in an "increasingly effective combination of digital telecommunication networks (the nerves), ubiquitously embedded intelligence (the brains), sensors and tags (the sensory organs), and software (the knowledge and cognitive competence)" (Chourabi et al., 2012: 2290).

Smart cities and the digital networks that link them together are best understood as emergent automation systems supported by interdependent subsystems of scaled technological and human intelligence. Where industrial cities were simply "skeleton and skin," the smart city is envisioned as a living organism containing "an artificial nervous system" (Chourabi et al., 2012: 2290). Building on layers of fixed Internet protocol networks and wireless satellite and mobile networks, smart cities are designed to leverage massive amounts of data generated by billions of Internet and mobiles devices and services. This includes:

1 Machine-to-machine (M2M) communication across mobile devices
2 Large-scale data processing via "Cloud Computing" in the processing and display of data
3 Data analytics and "Big Data" that correlate and interpret flows of knowledge and information

One of the key challenges confronting the ideology and theory of the smart city, however, is its *top-down* design. This includes a long-standing critique of the outsized influence of multinational corporations on contemporary urbanization. Indeed, the very idea of smart cities has been criticized for putting an excessive weight on economic values as the sole driver of urban development. Sassen (1996: 210), for example, calls attention to the ways in which neoliberal policies have used cities to concentrate and manage capital accumulation. As she observes, cities have become focused expressions of extreme inequality, marginality, and discrimination. Moreover, Harvey (1973) makes the point that neoliberal urbanization has only exacerbated social hierarchies of race and class in the structuring of urban spaces.

Emerging in response to smart urbanization is a counternarrative in favor of open and loosely coupled coordination systems. Indeed, in the face of expanding digital surveillance systems (both public and private), questions have surfaced about the ethics of smart cities. This includes a growing interest in shifting the discussion on smart cities "from the promotion and administration of services to the liberal democratic governance of their applications" (Allwinkle and Cruickshank, 2011: 9). Greenfield (2013), for example, suggests that smart cities should be redesigned to leverage open and free data sharing in the context of a new and broader calculus on civic technologies.

Contradictions of the network society

When examining the contours of smart cities, one needs to consider questions related to citizen empowerment vis-à-vis a network society. Alongside questions of open data and increased transparency, for example, are new possibilities for strengthening the capacities of communities and stakeholders to have increased political agency. Beyond the era of patriarchal power structures fixed to command-and-control leadership, network systems now challenge us to rethink the institutions and practices that underlie older notions of representative democracy. What is becoming obvious is that the laterally scaling practices that emerge from distributed networks are forcing a change in the constitution and meaning of "government." This is important because cities are increasingly becoming embedded in post-Westphalian era in which city-to-city cooperation can often supersede state-to-state cooperation (Campbell, 2012).

As Castells (2000: 500) notes, networked social structures now constitute the social morphology of our time. Linked to an expanded notion of government accountability is an argument for the use of digital tools and technologies that can enhance participatory democracy (Obama, 2009; Osimo, 2008). When considering the shifting architecture of democracy through the medium of ICTs, we see that networks provide an important means to understand both changes in the practice of democracy and an expansion in the notion of participation. This allows for a deeper understanding of the possible convergence between the distributive logic of ICT networks and the *bottom-up* logic of democratic self-governance.

In truth this is about more than smart cities; it is about the democratic ecologies that engage social capital in the delicate building of community. In addition to smart technologies, we should add the growing importance of social capital as a key feature of smart cities in providing the connective tissue through which knowledge can be accessed, adapted, and shared (Putnam, 2002). Most important, this bridges discussions on transparency in policy making with the web of smart technologies that are now beginning to reshape the fabric of our cities. Put differently, the value of smart cities exists beyond globalized business models or university campus environments. The value of smart cities is found in their capacity to leverage scaled collaboration between citizens as agents of creativity and civic participation.

The real basis for smart cities, in other words, should begin with peoples and communities. Beyond questions of infrastructure and technology, the key to truly smart cities is an enhanced capacity to support social capital and social agency. This implies a kind of development and growth that is supported by engaged citizens, civic institutions, and a wide range of policy actors across a digitally mediated networked commons.

Building smart cities from the bottom up

Building on affordances in sensor technologies, data analysis, and urban design, new policies and planning have the potential to leverage newer and richer forms of democratic engagement. Indeed, the growth of peer-to-peer (P2P) networks that augment next-generation communication, data sharing, and value creation have in fact opened a wide array of new opportunities for bottom-up civic engagement across a range of public services. The obvious question is: What is the potential of smart cities with emerging economies to become platforms for bottom-up civic engagement in the context of next-generation communication, data sharing, and application development? In conjunction with issues related to power grids, transportation networks, and water distribution systems, there is an acute need to examine the potential of smart cities as "networked ecologies" for citizen empowerment. Neo-Schumpeterian theories on techno-economic transformation (Perez, 2002), for example, suggest that an ongoing evolution in the relationship between innovation and social practice is fomenting a paradigm shift in the design and management of cities (Table 1.1).

Many now argue that the affordances of digital technologies are remaking the very notion of democracy. As Castells (2009) observes, the rise of socially mediated ICTs has sparked new social movements that have the capacity to build multiscale networks across a wide spectrum of sociopolitical environments Indeed, O'Reilly (2013) makes the case that automation is preferable to government incompetence. Using sensor technologies and ICT networks to reinforce government oversight, he argues that smart technologies could begin to "reduce the amount of regulation while actually increasing the amount of oversight and production of desirable outcomes" (p. 293). In his view,

Table 1.1 Techno-economic paradigm shifts.

Technological Revolution	Paradigm	Year	Core Country	Initiating Revolution
First	The Industrial Revolution	1771	Britain	Arkwright's mill in Cromford
Second	Age of Steam and Railway	1829	Britain (spreading to Europe and U.S.)	The steam engine in the Liverpool-Manchester railway
Third	Age of Steel, Electricity and Heavy Engineering	1875	U.S. and Germany	The Carnegie Bessemer steel plant in Pittsburgh, Pennsylvania
Fourth	Age of Oil, the Automobile and Mass Production	1908	U.S. and later Europe	Ford Model-T in Detroit, Michigan
Fifth	The Information Revolution	1971	U.S. (spreading to Europe and Asia)	The Intel microprocessor in Santa Clara, California

Source: Based on Perez (2002).

> Revelation after revelation of bad behavior by big banks demonstrates that periodic bouts of enforcement aren't sufficient. Systemic malfeasance needs systemic regulation. It's time for government to enter the age of big data. Algorithmic regulation is an idea whose time has come. (p. 291)

Coining the term "algorithmic regulation" to describe technology-mediated enhancements to government, he argues that computer algorithms could entirely displace layers of public bureaucracy. Accordingly, government regulations "should be regarded in much the same way that programmers regard their code and algorithms. That is, as a constantly updated toolset to achieve the outcomes specified in the laws" (O'Reilly, 2013: 291).

Notwithstanding the potential for abuse, advancing government into the era of Big Data could be key to reducing systemic malfeasance within the systems and networks that now sustain ICT-driven societies. This begs the question: What social structures are necessary for managing algorithmic governance and what is the potential for exploitation in such a vast system of digital management and control?

As Noveck (2009) observes, democracy itself now faces a crisis of legitimacy. This is because the rationale for closed systems of decision making (in which citizen participation is confined to voting or interest group activism) belongs to a different era. Much as civic groups have begun using ICTs to leverage political activism and social awareness, Noveck argues that ICTs are emerging as a platform for new forms of collaborative governance. Beyond deliberative democracy or democracy as discourse (Habermas, 1981), she argues for democracy as collaborative practice. This includes granular microtasks that are made available to participatory collaboration. In her view,

> The ability to organize collective activity puts more power in the hands of individuals by making it possible for people to self-organize and form teams around a boundless variety of goals, interests, and skill sets. And technology can support the formation of larger and more complex teams than previously imaginable. (Noveck, 2009: 32)

Where deliberative democracy focuses on inputs, collaborative democracy focuses on outputs (Table 1.2).

Table 1.2 Comparing deliberative and collaborative democracy.

Deliberative Democracy	Collaborative Democracy
1. Focuses on diversity of viewpoints.	Focuses on diversity of skills.
2. Measures the quality of democracy based on procedural uniformity and equality of inputs.	Shifts the focus to the effectiveness of decision-making and outputs.
3. Requires an agenda for orderly discussion.	Breaks down a problem into component parts that can be parceled out and assigned to other members of the public and officials.
4. Debates problems at the abstract level before the implementation of the solution. Or discusses solutions after its already been decided.	Is enabled by collaborative decision-making all through the process.
5. Focuses on opinion formation and the general will (achieving consensus as an end).	Focuses on collaboration as a means to an end
6. Focuses on self-expression.	Focuses on participation.

Source: Noveck, 2009, pp. 39–40.

Reimagining smart cities in the network society

What makes the idea of collaborative democracy particularly important to (re)designing smart cities is the idea that the affordances of technology are changing the relationship between citizen engagement and public management systems. Beyond simple representation, for example, the question that emerges is whether ICTs provide citizens with a capacity to *coproduce* public systems of governance. In truth, outdated theories of participatory democracy are holding back the evolution of urban political institutions and practices. In contrast to neoliberal conceptions of urbanization, what is needed now is bottom-up governance that centers on negotiation in the context of participatory collaboration. Indeed, this is about more than civic action or "social resistance from below." It is about collaborative engagement in the process of governance itself.

Beyond the corporate control of smart cities there is a growing need to consider a new focus on network technologies as providing a new kind of civic platform. Borrowing language and discourse form the Open Source Movement (OSM), for example, Benkler (2006) argues that the rise of networked environments make possible a new modality of organizing production in the form of "commons-based peer production." As he suggests, the key to understanding peer production is that resources within networks are held in common. That is to say, they are collectively shared, managed and produced (Rifkin, 2014).

Although it may be true that ICTs play a pivotal role in enabling citizens to access knowledge, information, and data, it is also the case that the ways in which members of a community relate to one another is a key factor in whether knowledge will have value as a social good. What is obvious, for example, is that the expansion of Internet protocol networks across the world has made it much easier to join humans together in laterally scaled collaboration (Rifkin, 2014). One of the central contradictions emerging with the affordances of so-called network societies is that information is now easily reproducible – leading to a variety of freedom/control problems related to intellectual property (IP) regimes.

The main problem is that knowledge and information function differently from other commodities because there are essentially zero marginal costs to adding more users to information systems. As Open Source (OS) advocates maintain, knowledge and information are social goods. (Lessig, 2004; Stiglitz, 1999). In economic terms, knowledge is a *nonrival* good – meaning that one person's use does not preclude use by another.

While the goal of IP protection is to incentivize creativity and invention by rewarding individuals for their work, the rise of a network society has introduced an entirely different set of values and incentives (Leadbeater and Miller, 2004). Indeed, Bloom (2013) argues that systems and practices that leverage networked data are best suited to community cooperatives and the administration of knowledge and information as a commons. As he suggests, opening and standardizing data is only part of the challenge ahead. The real difficulty is designing systems that enable interoperability.

The main issue with an open source society is that free and open data can be costly to maintain and necessarily requires support and caretaking to avoid deterioration. In order to maintain interoperability, Bloom proposes the use of community cooperatives that can simultaneously align diverse perspectives and interests, while also ensuring shared responsibility. This translates as systems of cooperation that can use and aggregate pooled resources (skills, time, money, knowledge), while sustaining open source capacities. According to Bloom, a "community data co-op" should include three primary roles:

1. The organization and standardization of the creation/aggregation of data as a common pool
2. The facilitation of data circulation through an ecosystem of services (both private and public)
3. The open education of co-op members through direct access and engagement with the creation and aggregation of public data

By democratizing access to knowledge through the use of open licensing, for example, smart cities might empower communities to participate in the production and consumption of commons-driven public resources without limitation.

Toward commons-driven smart cities

While it may be true that ICTs are introducing a range of new capacities for the design and planning of human urbanization, questions remain about the network capabilities of smart cities to remake systems of government. Beyond the simple transmission of public services, government systems are now increasingly under pressure to develop institutional frameworks that support tools and resources for empowering citizen agency. The overarching idea is that smart city planning should – at the very least – incorporate experimentation with new forms of community and citizenship.

Indeed, overlapping growing calls for open data and citizen empowerment is a rising demand for *Open Government.* Notwithstanding the fact that the concept of Open Government dates back to the European Enlightenment, the ideal of Open Government has become increasingly joined to contemporary debates on democratic reform. Widely viewed as a foundation to enhancing democratic governance, Open Government is closely tied to democratic reform movements and an expanded notion of citizen participation (Goldstein and Dyson, 2013; Lathrop and Ruma, 2010; Noveck, 2009). Claims supporting the value of Open Government suggest that citizen participation reduces government corruption by expanding public scrutiny and decentralizing government power. This overlaps European Enlightenment notions of free inquiry and free expression of opinion in the context of social equity and self-governance.

Building on an expanding literature on Open Government, there is increasing interest in reforming the practices and institutions that now define modern democracies. This includes increased advocacy for greater openness and greater transparency in political decision making and the reform of public service provisions. In the United States, for example, the Open Government directive introduced by President Obama (2009) has its foundation in regulations such as the Freedom of Information Act, the Paperwork Reduction Act, and the e-Government Act (McDermott, 2010).

What is becoming clear is that the revolutionary advance of ICTs has begun to overturn the older centralized logic of government. Expanding on the social capital that emerges from interconnected communities and a growing demand for citizen participation in decision making, governance is increasingly becoming about the redistribution and sharing of political power. Beyond the proprietary knowledge anchored to closed hierarchies of government, Open Government could well offer networked communities to enlarge citizen participation in the building of collaborative democracies for the 21st century.

What is also clear is that the ways in which public data are procured, administered, and aggregated should be as much a part of the conversation on smart cities as the final use of that data. As Gorbis (2013) argues, the network-driven dynamics that are reshaping industrial societies are reshaping governance itself. Like Noveck, she defines this as a shift toward a "new form of value creation that involves microcontributions from large networks of people utilizing social tools and technologies to create a new kind of wealth" (p. 25). This includes health, education, and financial data that can support modeling and simulations for better-informed decision making.

Underlying this reimagining of governance is an expansion in the tools and technologies that now support human decision making. Gorbis describes this as "socialstructing" and argues that the "growing accessibility and capability of tools for networked collaboration" is enabling users and citizens to unbundle many large tasks and distribute these tasks ad hoc across a community (p. 117). Put simply, as computational technologies continue to advance, smart systems will make it increasingly easier for average citizens to access real-time social and economic data. In the context of Open Government, for example, this includes the following (p. 102):

1 Rich and open data for making informed decisions
2 Sophisticated decision-support tools for exploring alternatives and uncovering complex interdependencies
3 Engagement platforms for wide citizenship involvement and deliberation
4 Microparticipation of regular citizens in government decisions and delivery of public services

Building on top of distributed networks, socialstructed systems could leverage the amplified power of individuals in collaboration around deinstitutionalized production and value creation.

Much as the goal of Open Government in a digital era involves rethinking notions of citizenship, so must the civic technologies that underlie notions of smart cities become more closely linked to new forms of commons-based social practice. This means moving past questions of government transparency and toward new modes of collaborative democracy.

Conclusion

Technology and democracy have been intimately linked throughout the modern period and yet today this linkage seems vastly more complicated than ever before. Indeed, the rise of the digital surveillance state reflects a new political system on the horizon that is testament to the need for rethinking the institutions and practices that oversee and govern modern democratic societies. At the same time, the rapid urbanization of human societies in the 21st century demands scalable solutions that empower network societies. The power of networks to provide vast horizontal scale offers both a way forward beyond the surveillance state, and a window into the democratic ecologies that we are only now beginning to perceive.

While ICTs are critical to serving as platforms for smart urbanization, it is citizens themselves who will solve – or not solve – the social, political and economic challenges we now face. As networks remake our society and governance structures, ever greater possibility exists for shaping smart cities as democratic ecologies. As this chapter suggests, the challenge for democratic governance in the context of smart cities is less about finding new solutions to the transmission of government services, and more about empowering citizens and communities to become agents in their own governance. The more citizens are empowered to engage as communities of practice in the generation of government, the more likely they will become genuine agents of community building in service to collaborative democracy.

References

Allwinkle, S., and Cruickshank, P. (2011). Creating smarter cities: An overview. *Journal of urban technology*, 18(2), 1–16.

Benkler, Y. (2006). *The wealth of networks: How social production transforms markets and freedom*. New Haven, CT: Yale University Press.

Bloom, G. (2013). Towards a community data commons. In B. Goldstein and L. Dyson (eds.), *Beyond transparency: Open data and the future of civic innovation*, pp. 255–270. San Francisco: Code for America Press.

Campbell, T. (2012). *Beyond smart cities: How cities network, learn, and innovate*. New York: Routledge.

Castells, M. (2000). *The rise of the network society, the information age: Economy, society, and culture* (2nd ed.). New York: Oxford University Press.

Castells, M. (2009). *Communication power*. New York: Oxford University Press.

Chourabi, H., Nam, T., Walker, S., Gil-García, J. R., Mellouli, S., Nahon, K., Pardo, T. A., and Scholl, H. J. (2012). Understanding smart cities: An integrative framework. In Proceedings of the 45th Hawaii International Conference on System Sciences (HICCS 2012), Maui, HI, January, pp. 2289–2297.

Florida, R. (2002). *The rise of the creative class: And how it's transforming work, leisure, community and everyday life*. New York: Basic Books.

Goldstein, B., and Dyson, L. (Eds.). (2013). *Beyond transparency: Open data and the future of civic innovation*. San Francisco: Code for America Press.

Gorbis, M. (2013). *The nature of the future: Dispatches from the socialstructed world*. New York: Free Press.

Greenfield, A. (2006). *Everyware: the dawning age of ubiquitous computing*. Berkeley, CA: New Riders.

Greenfield, A. (2013). *Against the smart city*. New York: Do Projects.

Habermas, J. (1981). *The theory of communicative action*. London: Beacon Press.

Harvey, D. (1973). *Social justice and the city*. Baltimore: Johns Hopkins University Press.

Hollands, R. (2008). Will the real smart city stand up? Creative, progressive, or just entrepreneurial? *City*, 12(3), 303–320.

Kanter, R. M., and Litlow, S. S. (2009). Informed and interconnected: A manifesto for smarter cities. Working Paper 09-141, Harvard Business School.

Lathrop, D., and Ruma, L. (Eds.). (2010). *Open government: Collaboration, transparency, and participation in practice*. Sebastopol, CA: O'Reilly Media.

Leadbeater, C., and Miller, P. (2004). *The pro-am revolution: How enthusiasts are changing our economy and society*. London: Demos.

Lessig, L. (2004). *Free culture: How big media uses technology and the law to lock down culture and control creativity*. New York: Penguin.

McDermott, P. (2010). Building Open Government. *Government Information Quarterly*, 27, 401–413.

Mitchell, W. (2007). Intelligent cities. *e-Journal on the Knowledge Society*. Retrieved from: http://www.uoc.edu/uocpapers/5/dt/eng/mitchell.pdf.

Noveck, B. S. (2009). *Wikigovernment: How technology can make government better, democracy stronger, and citizens more powerful*. Washington, DC: Brookings Institution.

Obama, B. (2009). Memorandum for the heads of executive departments and agencies – subject: Transparency and Open Government. Washington, DC: White House. Retrieved from: http://www.whitehouse.gov/the_press_office/TransparencyandOpenGovernment

O'Reilly, T. (2010). Government as a platform. In D. Lathrop and L. Ruma (eds.), *Open government: Collaboration, transparency, and participation in practice*, pp. 11–39. Sebastopol, CA: O'Reilly Media.

O'Reilly, T. (2013). Open data and algorithmic regulation. In B. Goldstein and L. Dyson (eds.), *Beyond transparency: Open data and the future of civic innovation*, pp. 289–300. San Francisco: Code for America Press.

Osimo, D. (2008). *Web 2.0 in government: Why and how?* JRC Scientific and Technical Reports No. EUR 23358. Luxemburg: Office for Official Publications of the European Communities. Retrieved from: http://ftp.jrc.es/EURdoc/JRC45269.pdf

Perez, C. (2002). *Technological revolutions and financial capital: The dynamics of bubbles and golden ages*. Cheltenham, UK: Edward Elgar.

Porter, M. E. (1990). *The competitive advantage of nations*. New York: Free Press.

Putnam, Robert D. (2002). *Democracies in flux: The evolution of social capital in contemporary society*. Oxford: Oxford University Press.

Rifkin, J. (2011). *The third industrial revolution: How lateral power is transforming energy, the economy, and the world*. New York: Palgrave Macmillan.

Rifkin, J. (2014). *The zero marginal cost society: The Internet of things, the collaborative commons, and the eclipse of capitalism*. New York: Palgrave Macmillan.

Sassen, S. (1996). Whose city is it? Globalization and the formation of new claims. *Public Culture*, 8(2), 205–223.

Stiglitz, J. (1999). Knowledge as a global public good. In I. Kaul, I. Grunberg, and M. A. Stern (eds.), *Global public goods: International cooperation in the 21st century*, pp. 308–325. New York: Oxford University Press.

United Nations. (2014). *World urbanization prospects: The 2014 revision*. New York: Department of Economic and Social Affairs, United Nations.

2

Government's Role in Growing a Smart City

Matthew Claudel, Alice Birolo, and Carlo Ratti

Growing the smart city

Our planet is urbanizing at a staggering rate: more than half the human population live in cities today, and the number is growing.[1] Today's urban space is changing rapidly, as digital technologies and pervasive networks integrate with physical space. "Ubiquitous computing names the third wave in computing, just now beginning," once noted Mark Weiser, Xerox Parc pioneer. "First were mainframes, each shared by lots of people. Now we are in the personal computing era, person and machine staring uneasily at each other across the desktop. Next comes ubiquitous computing, or the age of calm technology, when technology recedes into the background of our lives."[2] Ubiquitous computing, with its so-called Internet of Things[3] corollary, is creating a new urban condition: the smart city.

It is widely thought that smart cities have the capacity to respond better to their inhabitants and their environment, becoming efficient, sustainable and livable ecosystems. A number of books,[4] articles[5] and studies[6] have supported this claim. With the goal of smart urban optimization, a broad spectrum of implementation models are emerging in different parts of the world. But what is the role of government in the process of implementing smart city developments? How can smart city funding be used most effectively, specifically to promote innovation? And are huge sums of public money ultimately the right stimulant of smart cities after all?

Models for smart city government

Diametric approaches are appearing between the United States and, broadly speaking, of the rest of the world. In South America, Asia, and Europe, all levels of government are quickly identifying the potential

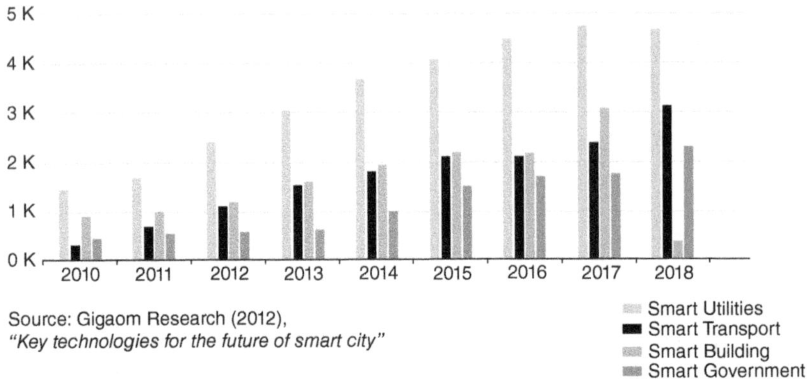

Figure 2.1 Annual Smart-City Investment by Industry World Market, 2010–2020.

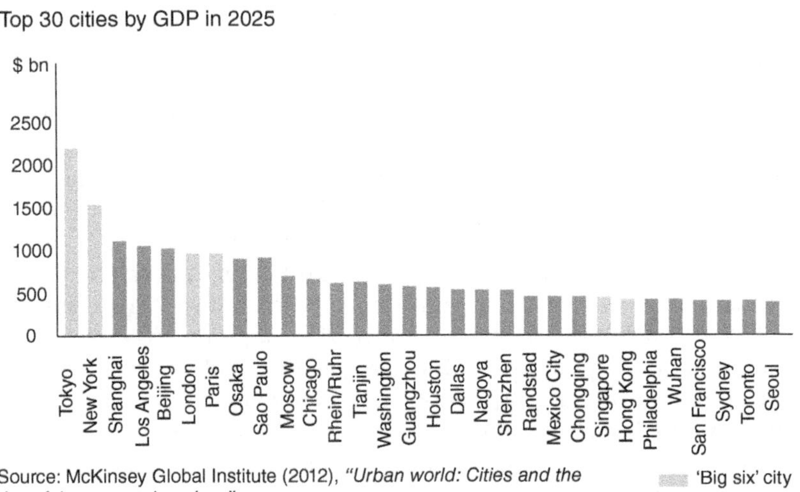

Figure 2.2 Top 30 cities by GDP in 2025.

latent smart cities, and are working to channel significant investment in that direction. Rio de Janeiro is building capacity at its "Smart Operations" center[7]; Singapore is about to embark in an ambitious "Smart Nation" effort[8]; and Amsterdam recently channeled €60 million ($81 million) into a new urban innovation center called Amsterdam Metropolitan Solutions[9]. The European Union's Horizon 2020 program has earmarked

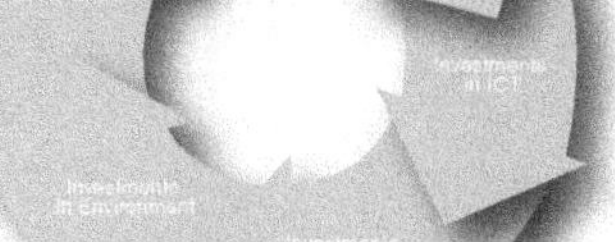

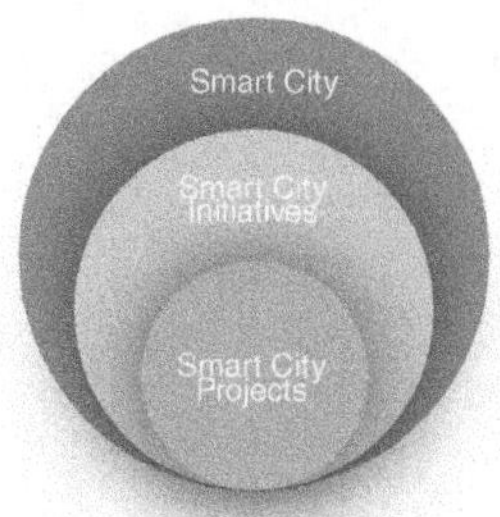

Figure 2.3 What defines a Smart City.

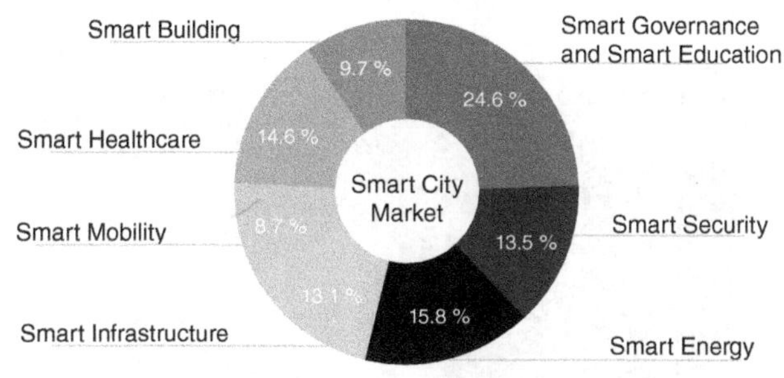

Figure 2.4 Smart City Market by Segments Global, 2012–2020.

€15 billion in 2014–2016[10] – an investment that represents a significant commitment of European resources to research and development in the field of smart cities, particularly during a time of severe fiscal constraints.

In the United States, on the other hand, there is little public sector funding, yet the general idea of smart urban space has been central to the current generation of successful start-ups. One recent example is Uber: a smartphone app that lets anyone call a cab or be a driver. The company's operations are polarizing: Uber has been the subject of protests and

Smart City Market
Most Adopted Funding Mechanism for Smart City Projects

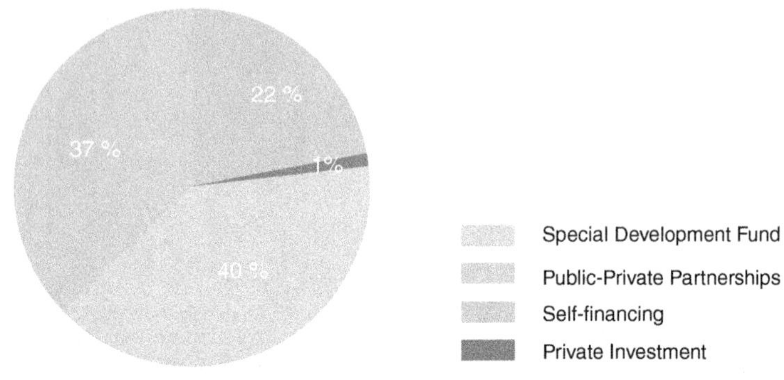

Source: Frost & Sullivan (2014), *"Strategic Opportunity Analysis of the Global Smart City Market"*

Figure 2.5 Smart City Market
Most Adopted Funding Mechanism for Smart City Projects.

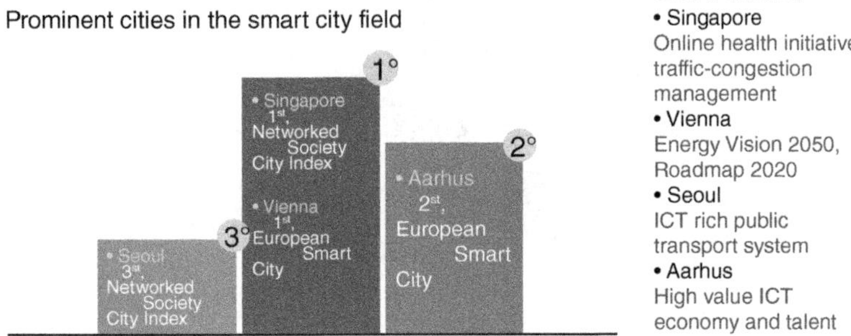

source: Jones Lang LaSalle (2013), *"The Business of Cities 2013. What do 150 city indexes and benchmarking studies tell us about the urban world in 2013?"*

Figure 2.6 Prominent cities in the smart city field.

strikes around the world (mainly in Europe), yet it was recently valued at a stratospheric $18 billion.[11] Beyond Uber, the learning thermostat Nest, the apartment-sharing website Airbnb, and the "home operating system" by Apple, to name a few, attest to the new frontiers of digital

Investments in smart city field
in American Cities

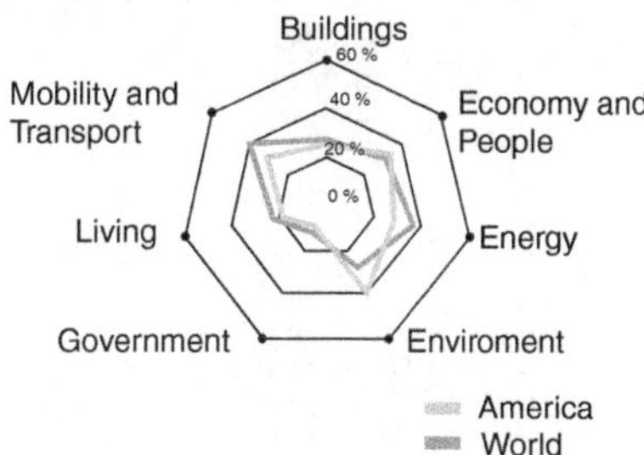

Investments in smart city field
in European Cities

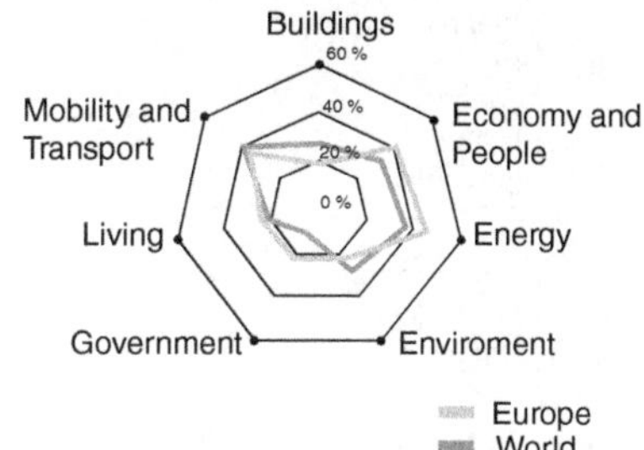

Investments in smart city field
in Asian Cities

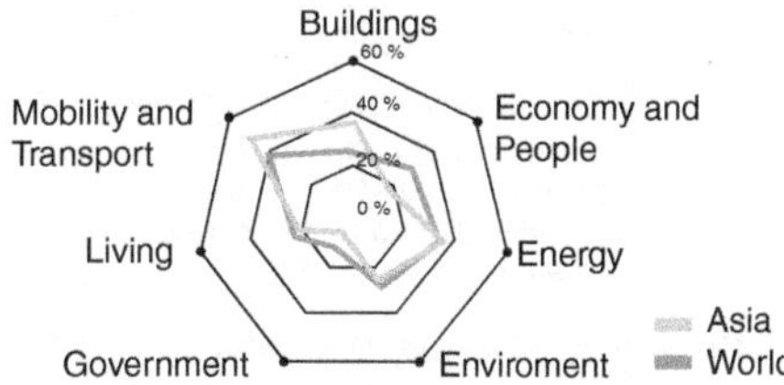

source: Cassa Depositi e Prestiti (2013),
"*Smart City. Progetti di sviluppo e strumenti di finanziamento*"

Figure 2.7 Investments in smart city field in American Cities; Investments in smart city field in European Cities; Investments in smart city field in Asian Cities.

information when it inhabits physical space. Similar approaches now promise to revolutionize most aspects of urban life – from commuting to energy consumption to personal health – and as such, they are receiving eager support from venture capital funds.[12]

That isn't to say that government should take a hands-off approach to urban development – it certainly has an important role to play. This includes supporting academic research and promoting applications in fields that might be less appealing to venture capital – unglamorous but nonetheless crucial domains such as municipal waste or water services. The public sector can also promote the use of open platforms and standards in such projects, which would speed up adoption in cities worldwide. Barcelona has made a step in this direction by creating the City Protocol[13]

Top 10 Smart Cities all over the world

Rank	City
1	Vienna
2	Paris
3	Toronto
4	New York
5	London
6	Tokyo
7	Berlin
8	Copenhagen
9	Hong Kong
10	Barcelona

source: Jones Lang LaSalle (2013), *"The Business of Cities 2013. What do 150 city indexes and benchmarking studies tell us about the urban world in 2013?"*

Best Quality of Life

Global Rank 2012	City	Global Rank 2011
1	Singapore	1
2	Sydney	1
3	Adelaide	4
3	Brisbane	3
5	Kobe	4
6	Perth	6
7	Canberra	9
8	Dublin	12
8	Melburne	10
8	Copenhagen	10
11	Bern	17
11	Hong Kong	14
11	Vancouver	14
11	Auckland	12
15	Antwerp	17
15	Wellington	14
17	San Francisco	20
17	Tokyo	6
17	Yokohama	6
20	Amsterdan	22

source: Jones Lang LaSalle (2013), *"The Business of Cities 2013. What do 150 city indexes and benchmarking studies tell us about the urban world in 2013?"*

Figure 2.8 Top 10 Smart Cities all over the world; Best quality of life.

that brings together cities, commercial and nonprofit organizations, universities and research institutions to develop a shared and interoperable set of guidelines and solutions for city transformation. Most important, these protocols will be multicity, multiculture, and multipartner.

But all of this is working toward less top-down determinism; governments should use their funds to develop an organic innovation ecosystem geared toward smart cities, similar to the one that is growing in the United States. It is more about bottom-up innovation than top-down projects. This must go beyond supporting traditional incubators, and aim to produce and nurture the regulatory frameworks that allow innovations to thrive. Considering the legal hurdles that continuously plague applications like Uber or Airbnb (paradoxically, Barcelona has been one of the most aggressive Airbnb opponents, fining the company €30,000 for tourism law infraction),[14] this level of support is sorely needed. Regulation is still vitally important, but in a more responsive way – government can still take the pulse of innovation and its impact on society, without creating unnecessary legislative constraints. Governments will have to be nimble on their feet, responding to technologies

2013 European Smart City Rankings

Final Rank	City	Smart Economy	Smart Enviroment	Smart Governance	Smart Living	Smart Mobility	Smart People
1	Copenhagen	7	1	7	2	4	1
2	Amsterdam	6	4	9	4	1	2
3	Vienna	4	6	3	1	6	7
4	Barcelona	5	5	5	6	3	5
5	Paris	3	7	8	9	2	4
6	Stockholm	8	2	4	7	7	6
7	London	1	10	7	10	10	3
8	Hamburg	8	3	10	3	5	8
9	Berlin	2	8	6	5	8	10
10	Helsinki	10	9	1	8	9	9

source: Co.Exist - World changing ideas and innovation (2013), *"The 10 Smartest European Cities"*

2013 Asian Smart City Rankings

Final Rank	City	Smart Economy	Smart Enviroment	Smart Governance	Smart Living	Smart Mobility	Smart People
1	Seuol	2	6	1	6	3	1
2	Singapore	4	1	2	5	2	7
3	Tokyo	4	2	5	4	4	3
4	Hong Kong	1	5	3	9	1	4
5	Auckland	8	8	4	1	10	2
6	Sydney	3	10	6	3	7	5
7	Melbourne	7	7	7	2	8	6
8	Osaka	9	3	8	10	5	9
9	Kobe	9	4	8	8	6	10
10	Perth	6	9	8	7	9	8

source: Co.Exist - World changing ideas and innovation (2013), *"The 10 Smartest Asia/Pacific Cities"*

Figure 2.9.1 2013 European Smart City Rankings; 2013 Asian Smart City Rankings.

as they emerge. In this manner, new developments will have room to grow, but their rise will be within the bounds of equitable operations.

The case of Singapore Smart Nation

Singapore is an apt case study, with the announcement of its Smart Nation Project (SNP), part of the government's Infocomm Media Masterplan. "Our goal is to establish Singapore as a smart nation that taps the potential of Infocomm and Media (ICM), and that nurtures innovative talent and enterprises," said Dr. Yaacob Ibrahim, Minister for Communications and Information. "In this way, the ICM sectors can bring about economic growth and social cohesion, and better living for our people."[15] But how? The city-state finds itself at a fork in the road toward smart city development, as the island becomes networked and intelligent.

2013 North American Smart City Rankings

Final Rank	City	Smart Economy	Smart Enviroment	Smart Governance	Smart Living	Smart Mobility	Smart People
1	Seattle	1	6	1	5	8	2
2	Boston	2	3	6	7	4	4
3	San Francisco	5	1	7	3	7	3
4	Washington, DC	10	4	2	10	2	1
5	New York	4	5	3	8	1	10
6	Toronto	2	9	3	3	6	9
7	Vancouver	7	2	9	1	9	6
8	Portland	6	7	5	2	10	5
9	Chicago	9	7	7	8	5	7
10	Montreal	8	10	10	6	2	7

source: Co.Exist - World changing ideas and innovation (2013), *"The 10 Smartest Cities In North America"*

2013 Latin American Smart City Rankings

Final Rank	City	Smart Economy	Smart Enviroment	Smart Governance	Smart Living	Smart Mobility	Smart People
1	Santiago	1	5	5	2	3	4
2	Mexico City	2	6	1	10	1	1
3	Bogota	4	2	3	9	4	2
4	Buenos Aires	8	7	2	4	7	3
5	Rio De Janeiro	3	3	6	7	8	5
6	Curitiba	7	1	10	3	6	8
7	Medellin	6	8	9	6	2	7
8	Montevideo	9	4	7	1	10	9
9	Lima	5	10	4	8	9	6
10	Quito	10	9	8	5	5	10

source: Co.Exist - World changing ideas and innovation (2013), *"The 10 Smartest Cities In Latin America"*

Figure 2.9.2 2013 North American Smart City Rankings; 2013 Latin American Smart City Rankings.

The first phase of the SNP will focus on the deployment of hard infrastructure, related specifically to connectivity and sensors, followed by initiatives that address various dimensions of the island's life and operations. These technologies constitute a "city operating system" similar to the software systems that run most of today's smart technologies, from laptops and iPads to increasingly networked domestic appliances.

While the masterplan spans the whole island nation, the Jurong Lake District development will become the heart of research and application, serving as a controlled trial area where smart city technologies can be deployed, tested, and subsequently transplanted elsewhere in the city (or across the planet). It will become an applied research site commonly known as an "urban living lab" in smart city jargon.

The goals of the SNP are ambitious. First and foremost is a concerted push for urban efficiency. Second, the plan seeks to promote an

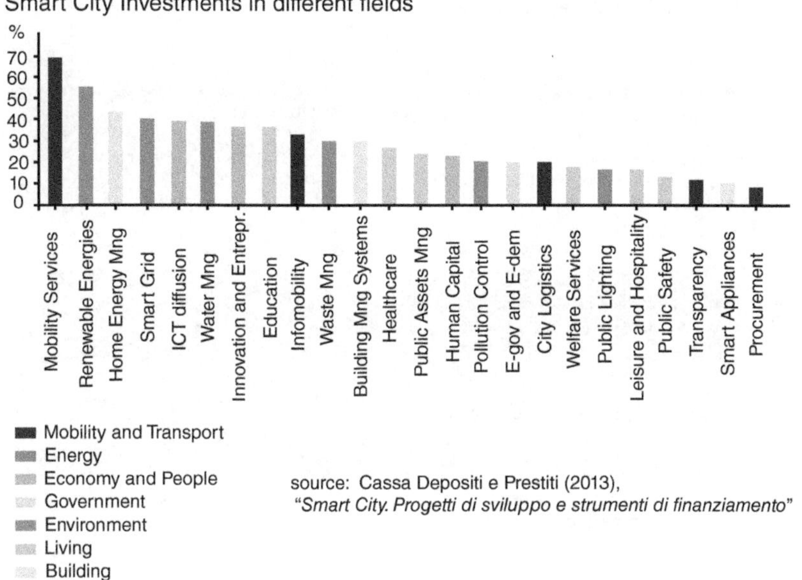

Figure 2.10 Smart City investments in different fields.

ecosystem of entrepreneurial innovation. Are these two objectives – efficiency and innovation – attainable? But most important, are they desirable?

The first goal of efficiency is quantifiable, and strategies for optimizing the city's function have the potential to make a substantial impact on daily life. Who would not want to live in a city that consumes less energy, or where traffic jams are reduced to a minimum? Singapore is probably one of the world's best test cases for cutting-edge urban developments. The nation is small, dense, tech- savvy, and most important, can now draw on an overt commitment from the government. This attitude is not new – it has transformed Singapore repeatedly since it became independent.

Transportation has been a recurring focus: Singapore pioneered one of the world's first Electronic Road Pricing schemes, later implemented by cities elsewhere. The system dramatically reduced vehicle traffic on roads, alleviating congestion, primarily in the central business district during peak hours. The public transit system is no less a model of efficient operation: since its inauguration, it has been rated the best Asia-Pacific metro system and most technologically innovative metro.[16] It is also among the most resource-efficient transit networks in the world, as evaluated by the international Metro Rail Awards.

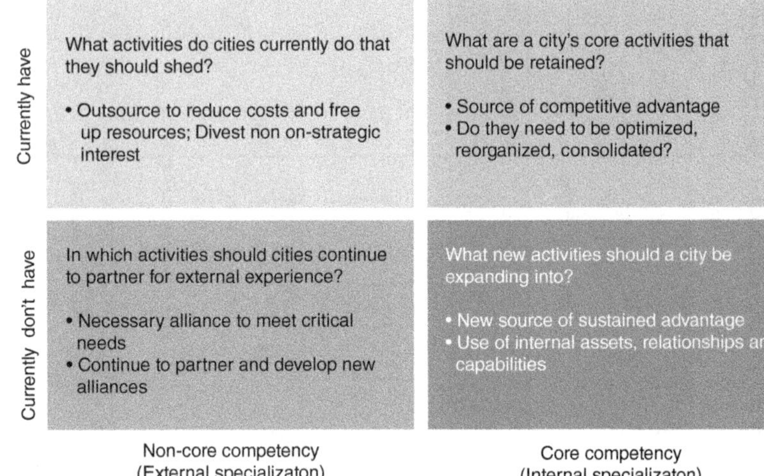

Source: IBM Center of Economic Development Analysis (2009), "*A vision of smarter cities. How cities can lead the way into a prosperous and sustainable future*"

Figure 2.11 A framework for strategic planning.

Today, car autonomy – as in driverless vehicles – is on the brink of entering the consumer marketplace, bringing significant benefits to society, drivers, and pedestrians. Singapore, once again, could become a world leader in testing future mobility. This is particularly promising in small controlled sites such as the Jurong Lake District or Sentosa, where autonomous driving projects have already been proposed.

But how will all of this spark innovation? Unlike efficiency, innovation cannot be institutionally purchased or mandated from the top down. It demands a complex and delicate ecosystem based on the bottom-up, concerted effort of many individuals. Here, Singapore's forward path will be more challenging. Mr. Lee Kuan Yew famously urged Singaporeans to take more risks, a vital component of the three attributes of global competitiveness underpinning his development platform: entrepreneurship, innovation, and management. "The American economy has taken off because of the enterprise culture and willingness to try," said Mr. Lee. "I think it's going to be a very arduous business changing the mindsets [of Singaporeans]."[17]

In the course of our work on the island, we have personally noticed this same pattern – government and business eagerly seek novel and

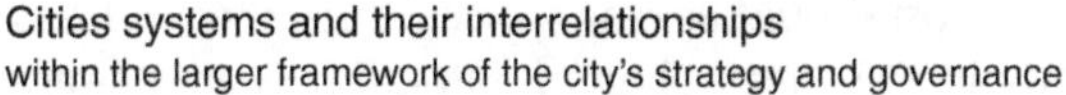

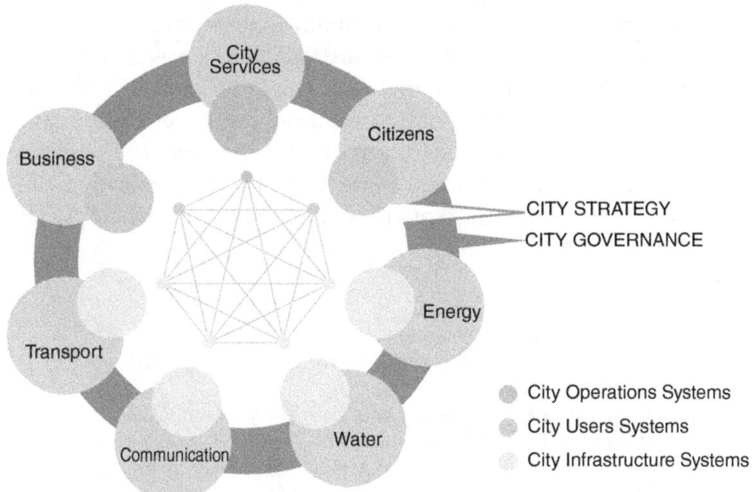

Source: IBM Center of Economic Development Analysis (2009), *"A vision of smarter cities. How cities can lead the way into a prosperous and sustainable future"*

Figure 2.12 Cities systems and their interrelationships within the larger framework of the city's strategy and governance.

innovative ideas at first, but soon furtively ask: "How many times has this been implemented before?" (By definition, if a technology has been implemented before, it is no longer novel!)

This is in sharp contrast to the prevailing attitude in California's Silicon Valley – one of the world's most productive innovation ecologies – where risk-taking is rewarded, while failure is tolerated.

Singapore needs this bold entrepreneurial spirit to exploit the cutting-edge tools that will be deployed in the course of the media master plan. Fostering an innovation culture will not be easy in a country where the educational system has historically been shaped by the stigma of failure [17]. Innovation demands an environment where ideas are tested and challenged, so that new and better ones can advance.

Innovation ecology

There seems to be a fine line for governments to walk as they implement smart city strategies: they should, at all costs, steer away from the temptation to play a deterministic and top-down role. It is not their prerogative to decide what the next smart city solution should be – or, worse, to use their citizens' money to bolster the foothold that technology multinationals are gaining in this field. Conversely, governments should create all the conditions – economical or normative – to grow innovation ecosystems.

And here might lie another delicate balance: between smart city efficiency and innovation. In some cases the latter will also need a good dose of chaos – the opposite of optimization – as the Singapore case study suggests. The most creative solutions often emerge and thrive in environments with less regulation and more mess. In other words, at times we might want less "smart" if "smart" is to be more than an empty label.

Notes

1 World Bank, Urban Development topic.
2 Weiser, M.. "Ubiquitous computing," http://www.ubiq.com/hypertext/weiser/UbiHome.html.
3 The term "Internet of Things" was coined by Kevin Ashton in 1999 for a presentation to Procter & Gamble (P&G) executives. However, the concept was detailed in a 1991 paper by Mark Weiser for *Scientific American*. "The computer for the 21st century." *Scientific American,* 265(3), pp. 94–104.
4 Glaeser, E. (2011). *Triumph of the city: How our greatest invention makes us richer, smarter, greener, healthier, and happier.* New York: Penguin Group, 2011.
5 O'Grady, Michael, and O'Hare, Gregory (2012). "How smart is your city?" *Science,* 335(6076) (March 30), pp. 1581–1582.
6 McKinsey & Company (2013). "How to make a city great." September.
7 Rio de Janeiro, IBM® Intelligent Operations Center, and IBM Smarter Cities Software. Inaugurated by Mayor Eduardo Paes in 2010.
8 Infocomm Development Authority (IDA) and the Singapore Smart Nation Programme, housed within the office of Singapore prime minister Lee Hsien Loong.
9 Amsterdam Metropolitan Solutions launched by former Amsterdam Vice-Mayor Carolien Gehrels.
10 Digital Agenda for Europe, a Europe 2020 Initiative. European Innovation Partnership (EIP) on Smart Cities and Communities.
11 Rusli, E. M., and MacMillan, D. (2014). "Uber gets an uber-valuation." *The Wall Street Journal,* June 6.
12 Herrmann, B. L., and Marmer, M. (2015). *Startup ecosystem report.* January.
13 City Protocol refers to both a program of activity and to the society that is responsible to develop this system's approach to rationalize, under a shared basis, city transformation.
14 Kassam, Ashifa (2014). "Airbnb fined €30,000 for illegal tourist lets in Barcelona." *The Guardian,* July.
15 Ibrahim, Yaacob (2014). "Speech by Dr Yaacob Ibrahim, Minister for Communications and Information, at the Opening Ceremony of imbX 2014." Infocomm Development Authority, Singapore, June.
16 Terrapinn (2010). Asia Pacific Rail Awards.
17 Ignatius, D., and Richardson, M. (2001). "Q & A/Lee Kuan Yew: Singapore's 'slow and painful' shift to Internet Age." *The New York Times,* January.

3
Generative Cities: Innovative, Sustainable, Inclusive

Ayesha and Parag Khanna

Introduction

In recent years, "smart cities" have become increasingly visible demonstration projects for showcasing sustainable architecture, low-energy power grids, connected transportation networks, and innovation clusters in new industry clusters. At the same time, as hundreds of major existing cities cope with rapid urbanization, aging infrastructure, limited finance, and rising inequality, many are seeking to retrofit, modernize, or upgrade various districts to become more livable and competitive.

Whether "old" cities or "new," a consensus is emerging that cities should strive to be economically innovative, environmentally sustainable, and socially inclusive. As the price of technology falls, sensors become increasingly ubiquitous, and data analytics widespread, what will increasingly differentiate cities is not how "smart" they are in terms of technology penetration, but the extent to which they leverage technology to meet these objectives. Technology is only part of the solution; policy matters just as much.

We believe that the concept of generativity can illuminate the ways in which people, technology, and policy interact in cities to achieve these progressive objectives, contributing to the ideal of cities becoming democratic ecosystems. In this brief essay we will discuss the concept of generativity, explore its normative implications, describe the ways in which cities can be platforms for advancing generative systems, and provide empirical evidence of efforts to leverage generative infrastructure toward more innovative, sustainable, and inclusive cities.

What is generativity?

Generativity is a broad property of systems that denotes the capacity of agents within them to connect to others and produce unanticipated outcomes and change. While the term's origins lie in psychoanalysis and linguistics, the Internet is now commonly understood to be a generative system. Jonathan Zittrain[1] of Harvard Law School writes that the Internet is generative because of its "capacity to produce unanticipated change through unfiltered contributions from broad and varied audiences." Indeed, the Internet is open to all participants, technically accessible to users producing code and content, and amenable to extension in unpredetermined ways. Such generative characteristics have enabled the Internet to become a kaleidoscope of applications created by a global community of users. Wikipedia serves as an example of how technological layers can enable generative content-production systems.

Today we can witness how technology is advancing the generativity of a wide range of social systems. From flip-teaching in the classroom to virtual currencies in the marketplace to citizen activist networks reshaping politics, human social organization is increasingly generative in nature. As some[2] have already observed, it is beginning to resemble the Internet itself.

A normative approach

Why is generativity important for the evolution of cities? Historically, great cities of the world have been characterized by strategic geographic location, demographic diversity, infrastructure quality, industrial innovation, vibrant culture, and global connectivity.

As urbanization rates rise around the world, however, rifts have emerged between cities and surround national areas, and within cities between core districts and peripheral or periurban areas. Indeed, the rapid acceleration of urbanization in recent decades correlates directly to the rise in income inequality within nations, even as it diminishes between them.[3] In the age of mega-cities featuring not only large populations but also great stratification of incomes and disparities of access to essential services, the extent to which all of a city's population shares in technological progress and its material benefits becomes an important qualifier as cities benchmark against and learn from each other.

Generativity thus becomes an important set of metrics, even an aspirational goal for cities that need to adjust to growing populations and rising demands for a better quality of life.

Advancing urban generativity

Generativity does not require new cities, but rather the pursuit of systems and policies that transform cities into innovative, sustainable, and inclusive communities. In this sense, cities become platforms, or living labs, of generative practices. Again, such practices must demonstrate both technological openness and also political commitment.

The technology platform that underpins generativity needs to be designed in such a way as to enable government efficiency and public access to useful data. This can include cloud computing services, sensor networks and data centers, and traffic management systems for both road congestion management as well as public transportation systems such as subways and light rail. Policies built on top of these platforms include e-government portals such as data.gov in the United States and other e-government services that allow citizens' access to data to shared application program interfaces (APIs) in order to create added-value programs.

For example, Code for America, a private initiative backed by major companies and foundations, trains dozens of fellows who embed in government agencies and small companies to optimize their usage of information technology. As U.S. state and municipal funding for information technology (IT) has risen beyond $60 billion (half the volume of U.S. federal spending), Code for America now has a special initiative for cities, growing from three city partners in 2011 to eleven in 2012, in each case expanding the range of data services and digitizing government request forms.[4]

Increasingly we see these experiments led from both the top down and bottom up. This is the essence of generativity, for connectivity is meant to empower citizens to pursue priorities in nonhierarchical ways. One pioneering example is Streetline, which allows commuters to leverage sensor networks to more efficiently find low-cost parking spaces. Streetline began as a private initiative, but would not have succeeded without linking motorists, parking providers, municipal transportation departments, and system integrators. Generativity then is about multidirectional and organic collaboration, leveraging technology in ways that inherently open up policy to broader participation. Saskia Sassen[5] has referred to this phenomenon as "open source urbanism."

Economic innovation

The essence of generativity is not to disrupt but to connect. While it is true that traditional economic relationships are being upended by new connections among producers and users of services through peer-to-peer (P2P) exchanges, at the same time, citizens and companies are also sharing collaborative spaces to promote mutually beneficial innovations. Both phenomena are part of the generativity trend. Examples of the former, such as Airbnb, have disintermediated the hotel industry. Examples of the latter are coworking spaces and government-subsidized innovation clusters such as Singapore's Block 71.

Together such social practices reduce costs to consumers while improving service quality and encouraging innovation. Firms such as Accenture and Plantronics have embedded teams in co-working spaces to better identify potential recruits and company acquisitions. Similarly, AT&T has created an in-house Foundry where start-ups are invited to work alongside the company's engineers. One of the start-ups acquired through this process helped cut the number of dropped calls on the network by 10 percent.

Economic generativity is also reshaping the urban workspace. As broadband Internet access spreads and the services share of the economy grows, an estimated three times as many workers will telecommute just one decade from now. Studies[6,7] show that it is more a lag in corporate culture that is holding back greater telecommuting rather than technological capacity. This example also shows how interconnected the digital environment is with the physical: rising remote and virtual employment can enable a virtuous circle of less commuter congestion just as Streetline does—and perhaps even more significantly.

Economic generativity also emerges from closer linkages among the educational and commercial domains. In cities, the physical master plan of efficient environments must be coupled with an economic master plan of labor force preparation and supply chain attraction. To this end, a growing number of vocational institutes have arisen in both developed and developing world cities, combining public funding with private expertise, to train thousands of potential employees and entrepreneurs in critical fields ranging from programming to construction management. For example, in Vietnam, where Intel has built a chip manufacturing facility employing 4,000 workers, the company has also launched a $7 million scholarship fund to train engineering students abroad and invest $2 million in supporting the creation of an in-country engineering master's degree program together with the Royal Melbourne Institute of Technology.

Environmental sustainability

Making cities more ecologically neutral can also benefit from a generative approach. Here there has been tremendous technological progress from zero emissions buildings in Seattle to carbon-neutral ports in Stockholm to traffic light management systems in Madrid. Los Angeles has replaced over 140,000 streetlights with LED technology, reducing energy consumption by 63 percent. These breakthroughs lay the foundation for more efficient public services and citizen interactions. In developing country cities such as São Paulo and Buenos Aires, there is now widespread emphasis on low-energy LED street lighting and low-carbon cement and other building materials used in the construction of commuter-friendly mixed-use commercial and residential real estate projects.

An important part of this shift toward sustainable infrastructure has been the rise of financial intermediaries such as infrastructure banks that take a long-term view and offer mechanisms to cover the high start-up costs of technology leapfrogging through covered bonds, credit risk guarantees, and cofunding with corporate financing arms. It is these new hybrid commercial arrangements as a form of policy innovation that can unlock technological potential across a far larger set of cities.

Sustainable infrastructure policy is not only about technology and finance but also citizen and consumer behavior. Generative cities create opportunities and incentives for people to participate in pro-sustainability practices. One example is Copenhagen's "bicycle superhighway," which is used year-round by people of all ages and socioeconomic status. In New York and Toronto, the designation of public space for carbon-offsetting tree planting has been coupled with broad social campaigns targeting up to 1 million new trees in the coming years. And in Beijing, engineers have attached sensors and lights to kites that measure and display air pollution levels to citizens below.

Other examples come from energy and water conservation. New York not only has mandatory energy audits for government, commercial, and residential buildings, but is also creating "solar maps" that allow residents to measure the solar power potential of buildings in which they live and work, presenting opportunities for cost savings and entrepreneurial innovation. Similar initiatives are under way to promote vertical farming projects that can boost the resilience of food supply, and the use of biomass for waste-to-energy power sources. As urban per capita water consumption grows, Singapore has distributed do-it-yourself water leakage repair kits to reduce water waste.

Transportation is a crucial new area of generative activity. The complexity of multimodal transport systems requires that a wide range of players from government agencies to car companies to advocacy groups come together. German cities have developed very strong frameworks and practices to transform their transportation systems in a manner making them more distributed and sustainable. Over a decade ago, Bremen added car- and bike-sharing to its mass transit system, and today Berlin's BeMobility program is introducing wide-scale electric car-sharing services.

Cloud services also empower increasingly mobile and urban populations. In Turkey, where each municipality is responsible for its own public transportation, transport software company Kentkart has deployed a cloud-based real-time General Transit Feed Specification (GTFS) system that allows all passengers with smartphones to view the live location and arrival times of buses and metros.

Social inclusiveness

The increasingly wide and deep technology penetration visible (and invisible) across particular cities and their residents through mobile phones and other wearable sensors has led some to argue that we are moving from the "Internet of Things" to the "Internet of People." This allows people not only to share information with each other in real time, but to also codetermine and act on each other's sentiments and priorities. Only through such civic engagement with technology can successful programs such as "See-Click-Fix" emerge across multiple American cities in which citizens respond to each other's inputs and problems as much as the government does. One sees such innovation in developing countries as well. One leading example is Bangalore-based Map Unity, a civic initiative to geo-locate not only transportation services, but also information about heritage sites, educational institutions, agricultural sites and prices, and health clinics. Leveraging real-time data from mobile phone towers, public bus GPS transceivers, and police traffic cameras, Map Unity effectively remapped Bangalore's bus routing to adapt to the actual community flow within the city.

Such dynamic services are crucial in urban slums, where one in five people in the world lives today. Slums require infrastructure, services, job creation, and other interventions that promote social and economic inclusion. In Mumbai, new housing is being developed to help shift residents of the city's largest slum, Dharavi, into permanent settlements. In Rio de Janeiro, new cable cars are in place to connect favelas to central districts, increasing both mobility and economic opportunity. Digital and people-based tools can be extremely useful in everything from mapping property rights to delivering mobile payments, legitimizing the otherwise informal transactions. All of these are examples of using connective technologies as enabling agents for greater social inclusiveness.

Conclusion

In this brief essay, we have attempted to raise questions that must always be at the forefront of conversations about smart cities: How transparent and cogoverned are new technologies deployed in urban environments? To what extent are innovation, sustainability, and inclusiveness strategically incorporated into new infrastructure investments? Ultimately, balancing the desire for control with the need for healthy chaos and experimentation are the essence of empowering a progressively generative city environment.

We must remember that generativity is a value-neutral property. Systems that are open to all can become not only vehicles for egalitarian policies but also monopolistic actors. From the prevalence of upgraded security cameras with facial recognition technologies in major cities such as London and Beijing to the fierce competition among "Silicon Superpowers" such as Apple, Google, Microsoft, and Facebook to dominate hardware, software, search engines, and consumer data, it is far from certain whether cities in the future will more resemble the "City of Control" or "City of Trust" from David Brin's noted 1998 novel *The Transparent Society*.[8] It is therefore most incumbent on the residents of generative cities themselves to harness their increasingly technological environment to shape urban life in directions that are innovative, sustainable, and inclusive.

Notes

1 Zittrain, J. (2009). *The future of the Internet—and how to stop it.* New Haven, CT: Yale University Press, p. 70.
2 Christakis, N. A., and Fowler, J. H. (2011). *Connected: The surprising power of our social networks and how they shape our lives—how your friends' friends' friends affect everything you feel, think, and do.* New York: Little, Brown.
3 Milanovic, B. (2012). "Global income inequality by the numbers: In history and now." *The World Bank.* Retrieved March 13, 2015, from http://elibrary.worldbank.org/doi/pdf/10.1596/1813-9450-6259.
4 2014. Partners. "Code for America." Retrieved March 13, 2015, from https://www.codeforamerica.org/governments/2014-cities/.

5 Sassen, S. (2011). *I bring open source urbanism and urbanizing technology* [Video]. Retrieved March 5, 2015, from http://postscapes.com/trackers/video/saskia-sassen-i-bring-open-source-urbanism-and-urbanizing-technology
6 Schadler, T., Brown, M., and Burnes, S. (2009). "US telecommuting forecast, 2009–2016—a digital home report." *Forrester.* Retrieved March 13, 2015, from https://www.forrester.com/US+Telecommuting+Forecast+2009+To+2016/fulltext/-/E-RES46635.
7 Maitland, A., and Thomson, P. (n.d.). *Future work: Changing organizational culture for the new world of work.* Retrieved March 13, 2015, from http://www.futureworkbook.com/.
8 Brin, D. (1999). *The transparent society: Will technology force us to choose between privacy and freedom?* New York: Perseus Books.

4

Urban Research Machines: Engaging the Modern Urban Citizen through Public Creativity

Anijo Punnen Mathew, IIT Institute of Design

When Rahm Emanuel was running for office of mayor of Chicago, he made an ambitious promise to the citizens of the city. If elected, he said, one of the first things he would do was work on a Cultural Plan for Chicago. Emanuel, being an economically savvy politician, understood the economic impact of culture on the city. Estimated to be around $2.75 billion, Chicago's cultural institutions along with other allied industries provide around 78,000 full-time equivalent jobs in the city.[1] Even before becoming mayor, he saw the Cultural Plan as a way to continue to elevate the city as a global destination for creativity, innovation, and excellence in the arts; a framework to guide this future cultural and economic growth. The last such plan was completed in 1984, and it established Chicago as a music, theater, and food destination. In 2011 when he was elected as the mayor, Emanuel combined two city departments into one, called it the Department of Cultural Affairs and Special Events (DCASE) and appointed Michelle Boone as the Commissioner of Culture. Almost immediately, Boone started groundwork on the Cultural Plan. She elected her officers, found a consultant, and started working on engagement opportunities. The 2012 Chicago Cultural Plan was soon under way—until she hit a wall.

Commissioner Boone knew that without the urban citizens' active participation, this plan would not gain traction in the city. She knew that no action can be built purely from the point of view of experts. But how could she reach out to 2.7 million people, and make sure everyone's voice was heard? The tools she had in her repertoire were extremely limited: either in the form of smaller scale face-to-face engagements such as town hall meetings, or extremely large-scale engagements such as surveys. While the modern urban audience is provided with many such opportunities to submit ideas and participate in discussions, participation is often very limited. There is no motivation for the urban citizen to stop in the middle of the road and answer a question about culture. Organizations that hold town hall meetings can often predict the demography that is likely to participate in these activities even before they take place. The bigger problem is that such engagements seldom capture the voice of the minority, and often ignore them for the voice of the people who are willing to participate. Commissioner Boone recognized this as a real problem, and she is not alone.

By the middle of this century, it is estimated that the global urban population will more than double, increasing from 3.4 billion in 2009 to 6.4 billion in 2050. The numbers are equally staggering in developing countries, changing from 2.5 billion in 2009 to almost 5.2 billion in 2050.[2] Designing for this new urban context becomes one of the most important agendas of our times. The modern urban dweller is a sophisticated and connected individual, for whom traditional models of community outreach may no longer be the only relevant form of civic engagement. Urban officials such as Commissioner Boone recognize that civic engagement is key to increased cooperation and collective action among the urban citizenry. Civic engagement opens up opportunities for the urban citizen to play an active role in the creation and sharing of information. An engaged citizen is one that plays a larger role in the improvement of cities and urban life. On a larger scale, these engagements are good because it allows organizations to understand the population better and respond almost immediately to the needs of the citizen. In the past, parts of the city were built specifically for such purpose. The commons, the marketplace, and the agora were all venues designed to reach out to reticent and yet important citizen voices in the city. The mere act of gathering in a public place demonstrated an urban citizen's preference or lack thereof. In recent times we have seen significant changes in people's sensibilities and expectations, especially with the coming of mobile, locative, and ubiquitous technologies that dominate every aspect of our lived experiences. New tools of social media create unprecedented opportunities to share, to cooperate with one another, and to take collective action. Yet, these digital tools are also changing the balance of participation and spectatorship among younger and more technologically savvy citizens of the city. As traditionally diverse communities become more socially connected through technologies such as Facebook, Twitter, Foursquare, and so forth, these connections will be the key element by which cultural inquiry can happen. The general population is getting used to the idea of having multiple outlets for their voice, but often at the cost of physical interactions and gathering. And those commons, marketplaces, and agoras that were once the center of a democracy? They are seldom used as gathering grounds anymore, at least not in the same way as before.

The question in the context of new urbanization, however, is what is the shape and form of the modern agora? How can cities reach out and actively listen to citizens that belong to a different generation altogether?

Emergence of a creative public

Leading up to the 20th century, information flow was controlled by those who had the ability to broadcast it. Publication houses chose the content that would populate the books they published; TV and radio stations broadcast shows that they decided would work; large production houses determined the type of movies that would be released into cinema halls. In all of this users on the ground for the most part consumed what was broadcast to them. Cities, too, reacted similarly – if they had to present an idea, the default mechanism was broadcast. Agoras transformed from a meeting places to a places where messages could be announced. By the middle of the 20th century this relationship shifted considerably. Shirky and others (Shirky 2009, 2010; Anderson 2006; Gee 2009) describe how that in the last few decades new tools of "social media" have enabled the everyday citizen with new powers. With the Internet, the public could now play a part in what kind of information was being published and broadcast. They could curate the content that was published for themselves as well as for others. They could also publish their own versions, maybe even create new types of information. Services emerged that responded to this demand, websites such as Digg and Reddit allowed users to curate existing content. MySpace allowed users to compose short blurbs of information that could be broadcast to many others. Facebook, Google Plus, YouTube, and Twitter allow the creation of information for specific groups. MeetUp and Foursquare allow people to broadcast place-based information, and services such as Yelp enables user to decide on where they eat or play based on reviews of everyday users like themselves. These digital tools are starting to play a significant role in the politics of modern governance. They not only open up an ability for everyday users to produce information but they also shift balance of participation. Erstwhile consumers of information are now also producers of information. So much so that traditional producers – publication houses, TV broadcasters, radio stations, and even governments – cannot keep up with the pace at which this information is produced.

This shift in balance can be explained in four ways:

1 It is easier today for everyday people not only to consume media but to also produce it themselves. The general public, not only experts, can produce content as much as they consume, using relatively simple tools and services.
2 Digital tools are also changing the balance of participation and spectatorship. People are no longer restricted to the role of the spectator. As participants, they can (and often do) participate in what used to be creative practices reserved for professionals.
3 Information is not fixed; it is being aggregated and constructed upon. So much so that it is often possible to take a single information source and watch it change over time because of content generated by users.
4 Digital tools are changing the nature of groups, social formations, and power. Today, with social networking sites like Twitter, Flickr, and Facebook, and digital devices like mobile phones, it is easier than ever to form and join groups, even for short-term purposes. Without the requirement of formal structure, groups can organize bottom up around specific topics or interest areas through constant communication and feedback. In fact what we are seeing today are new "tribal" relationships (McLuhan 1964/2003).

Such interactions are not just a behavioral attribute of users on the Internet, they are shaping the way our society deals with information in physical spaces as well. Some artists and urban designers are starting to use the city as a canvas to channel this creativity. Take Candy Chang, for example, an architect, urban planner, and designer who has designed several urban installations which combine social activism with place making. Chang's installations are simple, nontechnological, and often just a field of tags often just asking one question – "I want...on Broad Street" or "Before I die..." A famous series of installations called civic input on-site, places fill-in-the-blank stickers starting with the statement "I wish this was..." on the façade of vacant spaces in the city as an experiment to see what citizens would like to see in those spaces. The uniqueness of her work comes from the fact that she places these tags in provocative places such as an empty storefront in New Orleans, or a shuttered foreclosed home in Pilsen, Chicago. The end experience of the urban installation is generated not by Chang herself but from user-generated content, aggregated over a period of time (Chang 2012).

Chang cleverly engages proximal relationships – relationships between people established through co-location or proximity. And she does so by asking very simple questions about these places, forcing people to reflect and think about these relationships. The effect of these interactions do not come from one person answering this question. It comes from many people adding their answers or commentary to this simple question. The people who answer do not necessarily share a proximal relationship with each other; they share a relationship with the place. It is because of this relationship that they are willing to engage in a social conversation with others who share the same relationship. As more people respond, patterns emerge; as patterns emerge, responses change; as responses change, users react. In the beginning what looked like a field of empty tags is now filled with information from people on the ground. One individual tag may or may not be interesting, but put together the whole field starts to take on very interesting form. It becomes a call to action. Not just for the people who responded but also for the city, and for those in governance. In this manner, information completes an experience that design helped to initiate. And the experience becomes an enquiry into a community's feeling for place.

The deep engagement that Chang's installations encourage are a result of what I refer to as *public creativity*. *Public creativity* is defined here as collective or social (group) creativity which comes from the aggregation of user-generated content in *public space*. Chang's role as a designer is not to design the complete installation but to provide the levers by which users are able to construct their own experiences over time. The experiences do not arise merely from the artifacts that Chang places in the city but from collective or social (group) creativity. The voice of the installation is the voice of the participants – aggregated over time from publicly generated content. And more important, it allows designers to ground interaction in physical space, something that may be quite valuable in the time of hyper-mobile interactions everywhere.

Why physicality matters in the time of hyper-mobile interactions

Ito et al. (2008) in the John D. and Catherine T. MacArthur Foundation Reports on Digital Media and Learning highlight the social aspects of social media and its ability to enable peer-based learning. Their report point to two specific ways children use the Internet to learn – most youth, they claim, use online networks to extend the friendships that they navigate in the physical environments of school, religious organizations and activities like sports, and other local activities. These "friendship" networks are "always on," in constant contact with their friends via texting, instant messaging, mobile phones, and Internet connections. A smaller number of youth also use the online world to explore interests and find information that goes beyond what they have access to at school or in their local community. In these "interest" networks, youth may find new peers outside the boundaries of their local community. These two networks both exist at the same level, and yet they function at two different levels. One is a proximal relationship, established primarily through physical connections. The second is a social relationship, established through social activity built around shared interests, not necessarily proximal. Public creative installations such as Chang's are platforms where both of these relationships can potentially intersect. Almost all of the people who interact with the installation have some proximal relationship with the place, but they may also have similar relationships with others who interact as well. In either case, it is the physicality of the experience brings these people together.

On another front, new technologies and social practices are changing the nature of physical space itself. Mergers of physical and digital infrastructure mean that more information can be embedded into seemingly everyday objects in spaces, surfaces, and artefacts familiar to users and yet enable new meaning and creative experiences. The coming of ubiquitous computing signals a paradigm shift in computing research as described by Rogers (2006) – it means that computing can now be embedded into virtually any object, or environment and novel experiences can be conceived. Take, for example, Rafael Lozano-Hemmer, another artist-urban designer who engages *public creativity* in his projects, using large urban surfaces and technology.

In the Body Movies project, Lozano-Hemmer projected over 1,000 portraits using video mapping techniques on the façade of the Pathé Cinema building in Schouwburgplein square in Rotterdam. According to Lozano-Hemmer (2002) the coming together of physical infrastructure and virtual information means the user finds themselves uniquely positioned within the context of the urban space, realizing that they have the power to effect change on this space. Lozano-Hemmer offers a perspective about his work which is poignant to note here. He claims:

> *My work is best situated somewhere between architecture and the performing arts. For me it is a priority to create social experiences rather than to generate collectible objects. The making of a piece itself is closer to developing a performance or a play than a visual artwork.... You have this frame and you step back from the subject, from reality, as though looking through this neutral glass....* [In my work] *spectators play an active role, not a*

> *passive one. People who are participating are in fact reflecting. People are not innocent when they activate interactive works in a public space, and this already constitutes a certain ground for reflection. People are participating in these sort of interactive operations with a lot of knowledge and awareness.* (pp. 2, 3)

Lozano-Hemmer's work is a critique of place. The experiences constructed through the meeting of relative strangers in a large urban plaza enables reflection on the meaning of that particular place in the "here and now" of the interactions. Such reflection is key for civic engagement, because when centered on the immediacy of the problem or the context, discourse is much more meaningful and engaging. This is why physicality is important even in times of hyper-mobile interaction – on one hand, it enables connections based on proximal relationships in place, while on the other hand, it allows users to focus on the "here and now." Moreover, such engagements are designed to be never complete – the urban public continues to extend their engagement long after the installations themselves are removed. The end experience is derived from collective co-creation of content in physical space, the transformation of this content in near real time, and the reflection that arises from this transformation.

At IIT Institute of Design, we believe that such public creative installations have the potential to become agoras of new urban environments. Due to their computational core, they respond to the connected sensibilities of the modern urban citizen, while the physicality of the interaction maintains and nurtures proximal relationships. Over the course of the last few years, we have been developing both theory and prototypes to see if public creativity can be used to develop technology enabled urban interactions at the intersection of physical and digital spaces. The question we seek to answer through this work is, can public creative installations replace or augment traditional urban engagements?

Public creativity – a tool for urban engagement

The Chicago Loop Alliance is a Business Improvement District (BID) organization whose primary purpose (as defined on their website) is to develop, support, and promote artistic, cultural, and public events that benefit businesses, individuals, and stakeholders within the service area of the Chicago Loop. These initiatives enhance the character of the Loop, contribute to its competitive position as a mixed-use destination, and promote economic development and tourism in the area. In 2010, the organization approached the IIT Institute of Design (ID) to explore the idea of using public installations for engaging audiences on the street. As a business organization, they are constantly trying to understand the core demography of their district, and they wanted to explore interactive place making as a way to get to this information. Chicago Loop Alliance worked with ID to develop a comprehensive specification; they saw these projects as prototypes to explore new ways of communicating with its demography on the street. They also wanted to see how weaving technology into physical infrastructure, vis-à-vis interactive place making, could create a better ambience for the people on the front end and enable them to learn about their patrons on the back end.

This interaction led to a series of three interactive installations on historic State Street in Chicago. Two of these are presented here:

ZeroZero

The intersection of State and Madison is the center of the addressing scheme for the City of Chicago and has been so since 1909 when this new system was implemented. The addressing scheme uses a grid system with has a "primary" street at each half mile, and eight city blocks measure one mile and marked in increments of 100 from the origin of the grid at State (0 East) and Madison (0 North). Thus, one can easily see that Michigan Ave (100 East) is one block east of State St and Congress Parkway (500 South) is five blocks south of Madison St. Equally important is the fact that one of Louis Sullivan's most recognized designs, the Sullivan Center (or the Carson Pirie Scott building), is located at the southeast corner of this intersection.

For this project, the story of State and Madison became a story of recentering. It was about recentering around the city, its many neighborhoods, countless restaurants, bars, theaters, and playgrounds all of which can be traced from the city at State and Madison. It was also a story of recentering around the history of Chicago. State and Madison at one time used to be the busiest intersection in the world and even today has a large pedestrian traffic. The installation highlighted this by first giving State and Madison a personality – we called it "ZeroZero." Second, we did it by making visible the invisible stories of the countless people who pass along this intersection, a composite of the infinite number of place narratives that pass through the intersection.

The ZeroZero installation had two embodiments: a physical one and a virtual one. ZeroZero's physical embodiment was designed into the corner windows of the Sullivan Center. The window installation consists of a sculpture, an iconic world map, and instructions on the windows. ZeroZero's virtual manifestation[3] consists of a website and a mobile page. At the website, one could read about the history of State and Madison, and contribute to the place by adding an address anywhere in the world and describing why it is important to him/her. For example, I could add my home address and describe it as where I live; or add a University address and describe it as where I went to school; and so on. The website then uses a Google Maps API to pull an image from the address entered and add it to a database. The visualization pulls all of these images to show the many different stories of people who have

Figure 4.1 ZeroZero showing its two embodiments – a physical installation in the Sullivan Center and an online installation at zerozerochi.com.

interacted with ZeroZero – not just location, but also the place narratives. All inputs were anonymous, no identifying information was asked for or could be input. The system then translates the number of miles to each user's chosen location into a distance measured from State and Madison in terms of Chicago blocks (based on the fact that 1 Chicago block = 8 miles).

Upon installation, ZeroZero showed that people were willing to add their narratives to such interventions. We calculated that about 2,344,979 blocks were traveled at the end, and hundreds of place narratives collected. Clearly people wanted to tell their stories to others. The integrative power of the installation, however, comes from its back-end tracking. By tracking place information, CLA was able to track where people on State Street were coming from – not just through quantitative metrics (which students were able to derive from Google Maps) but also through qualitative metrics (through the descriptions that they add to the story they share). Moreover, no identifying information was asked for, no login was required, no names or identification data had to be provided during the interaction. Any information provided was purely voluntary and with the knowledge that it would be shared among others visiting the interface. CLA was given access to the data and shown how they could use it at both a macro aggregate level (were the users primarily tourists or residents?) or at a micro specific level (what makes certain places valuable for people?).

While ZeroZero took a direct approach to querying people, the second installation took a less direct route:

Urban Forest

Urban Forest embeds the concept of a social family tree into two windows of a large urban shopping mall called Block37. The location of Block37 has a storied history – it is famous for being perhaps the most prominent vacant lot in the country (Sharoff 2007), having remained vacant for over 20 years. When the sleek steel-and-glass urban mall, designed by the architecture firm Gensler, was finally constructed in 2007, it saw itself caught in the growing woes of the recession and the overall decline of large urban shopping centers – several of its high-profile clientele left and a superstation for an express route to the Chicago O'Hare airport disappeared because of lack of funding. The story of Block37 is a story of connects and disconnects between politics, urban infrastructure, and business. The site thus provided a perfect backdrop for the project which worked on the premise that everyone is interconnected in the city, often through relationships forged by an engagement with the city, sometimes in far greater ways than what is visible. Urban Forest is an interactive installation that attempts to highlight these relationships by making them visible – at the street level.

The final installation was identified by the CLA as the simplest and the most successful of the projects. The installation uses the two display windows to ask passers-by to answer one of three questions – thin crust or deep dish? Sox or Cubs? O'Hare or Midway? The questions were designed to change out every 72 hours. The interaction model is simple: as you walk down the street you tap on the question you associate with. The tap is visualized as a leaf on a digital tree that grows with every answer. As more people answer, the tree grows larger and larger, and the visualization allows passers-by to see which question is getting more responses from people on the street. So if you are passionate about deep dish pizza and you see that thin crust is winning, you can choose to add to the deep dish side. The one-step interaction model used (touch the question you want to answer) comes from embedded user research which suggested that passers-by like to note their preferences without hassles of log in or connecting using other mediated interfaces. This

Figure 4.2 Urban Forest showing the simple touch interaction and the visualization of the trees and the urban forest.

meant that passers-by were free to interact with the installation anywhere along the thresholds of activity framework (Brignull and Rogers 2003) – through peripheral awareness as they walked across the installation; through focal awareness as they watch others interact with the installation; or through direct interaction as they add their response to the "forest."

When installed on State Street, the numbers of responses for each question were tracked for two purposes – first to render past responses in the background as a "forest" constructed over time representing the collective preferences of a city; the second to construct a demographic profile of the population which frequents the urban location of Block37 on State Street (and the Loop). The second purpose for tracking is important as both Block37 and the CLA, the sponsoring organizations, were interested in increasing foot traffic in that area. The business premise was that understanding the demographic profile of the pedestrians on the street will allow both organizations to build experiences that can be catered to specific populations. This aspect of the project played a large role in the framing of the questions because aggregated responses to even a seemingly benign question such as "O'Hare or Midway?" (Chicago airports located at two ends of the city) could potentially lead to even abstract understanding of where the respondents come from.

As expected, CLA using all the aggregate data collected over the many weeks of installation was successfully able to build demographic models of people who inhabited State Street. Both ZeroZero and Urban Forest are *Urban Research Machines* – perfect examples of how interaction design can be used for urban enquiry.

What are Urban Research Machines?

Urban Research Machines reimagine how urban organizations can transition from traditional forms of user engagement such as surveys, focus groups, and town hall meetings, to more interactive and participatory interventions embedded in the heart of the city itself. Such installations can be located in population hubs around the city (and sometimes not located at all), and enable socially connected user engagement. One advantage of such machines is that they can be designed to serve as "constantly on" participatory environments for users to share, view, and build on each other's ideas. In other words, the urban audience are not just responding to questions they have been asked, but rather, reacting to others as much as they discuss their own preferences. *Public creativity* is the prime basis of interaction at these installations; the installations are designed specifically to allow the urban audience to "play" with information, and through such play, share information, cooperate with one another, and to take collective action about the topic at hand. Unlike a focus group meeting, these machines do not have an engagement time frame. Users can engage with them when they want, and where they want. This enables organizations to hear a collective voice aggregated over many weeks instead of singular voices in one sitting. Because some variations of such machines have no spatial embodiment or sometimes an extended spatial embodiment, users can engage with information from anywhere. Urban research machines are also designed to seek anonymous data – no identifying information is collected or stored. Information is often presented in a simple, playful manner which allows users of all ages and technological capability to engage with information.

Over the course of 2010–2011, we installed several such machines as prototypes in the city of Chicago. The place-making and civic engagement capabilities of these installations caught the attention of several government agencies in Chicago, Singapore, and Hong Kong. This is when Commissioner Boone heard about the machines installed on State Street.

Urban Research Machines for the 2012 Chicago Cultural Plan

In spring 2012, Commissioner Boone launched the first phase of the new cultural plan for Chicago. The focus of this phase was build collaborative engagements with public and private sectors, and most important, the civic community. For Boone, public engagement was a key element that would help shape Chicago's cultural future, so her team worked on several public community meetings, and an interactive website enabled Chicagoans to submit ideas and participate in a discussion about the city's cultural future. Based on the success of the State Street machines, the Boone team reached out to the IIT Institute of Design with a specific request. They wanted to explore a similar idea for raising awareness of the Cultural Plan and to reach out to the population. Over the course of several months, three ID teams worked with the DCASE to install similar Urban Research Machines in cultural hubs around selected neighborhoods. The installations are designed to monitor interaction and capture user information at these hubs. At the end of the project, both ID and DCASE hoped that the data generated would augment traditional forms of research, in particular by listening to the voice of a larger diversity of Chicagoans as they move through their everyday lives.

Three projects were deployed around the city – one in the City Hall within the Loop district, one at the historic Old Town School of Folk Music in the Lincoln Square neighborhood, and the last one in the Pilsen neighborhood in association with the National Museum of Mexican Art. Two of these are presented here.

SkyWords

SkyWords was designed to create awareness of and ask people about what parts of the city culture they found valuable. SkyWords is a site-specific installation created for the bustling ground floor of Chicago City Hall. One side of Chicago City Hall houses almost all of the important offices of the Chicago governance, while the other side houses many important parts of Cook County (the county Chicago is in). Scores of people pass through City Hall each day, and it is safe to say that the demographics mirror the demographics of the city itself. This cultural and linguistic diversity compelled us to find a simple, universal metaphor that everyone could relate to. The transience of the space and rushed attitude of the audience inspired us to create an interaction that was very easy to understand and fun to play for five seconds or five minutes, using balloons!

SkyWords is composed of two 8 x 4 x 12 foot installations positioned across from each other. One side of the installation asks questions based on Cultural Plan themes — for example, music, food, art, and community — asking people to share their preferences and aspirations for the city. Almost all of the 150 SkyWords questions were shaped by the Cultural Plan themes, which included education, participation, and resources. In order to generate useful data, the questions were constructed to elicit people's preferences and aspirations for Chicago's cultural life. For example, "Does your perfect Chicago have more hip-hop, jazz, or rock and roll?" To answer a question, the user presses the colored button and then — by pumping a bellows, spinning a pinwheel, or plunging a bike pump – physically! This action blows up a balloon containing the answer. Microcontrollers at the back of the installation read the user's physical interaction as mechanical input and ultimately translates that into a digital output – a balloon that rises on the screen. The more the users pump on the bellow, the bigger the balloon grows. Once the users release the balloon, it floats up slowly across the screen, almost as though it is floating up to the Mayor's office. On the other side of the installation, each inflated balloon that rose from the previous interaction descends onto the screen from the top. On this side, other users, using a joystick and good aim, can pop the balloons to reveal answers sealed inside. This back and forth was important as users not

Figure 4.3 SkyWords installation at the Chicago City Hall showing the two different parts and engagement models.

only input answers to questions, but they could also discover what other people said as well.

During SkyWords' 10-day run, hundreds of people used the installation; some played just once, others came back every day, many brought their friends and coworkers. On the first day of the installation itself the system received 750+ data points. We watched them pause to read the informational posters, take stacks of Cultural Plan flyers, and have conversations with strangers about favorite theaters and neighborhood festivals. SkyWords saw great success as an Urban Research Machine – people answered nearly 2,500 questions over the course of 10 days.

The second project was equally interesting, but took on a shape and form that was completely different from SkyWords:

The City Listens

The City Listens was located at the historic Old Town School of Folk Music in the Lincoln Square neighborhood of Chicago. It was built on the concept of public creativity, engaging users' creativity as a method to understand their opinion of art and culture. Playing off the air of performance and collaboration at the Old Town School of Folk Music, the installation allowed visitors to "talk" to the City by recording words, playing music, or otherwise expressing themselves through sound. Focusing on the Old Town School of Folk Music in Lincoln Square, we observed a rich variety of musicians, teachers, and performers of all ages. They were already telling us what culture meant to them, the city just needed to listen. In order to capture their stories, we asked people to share past memories, opinions about their present, and aspirations for the future. The City Listens opened a channel not only to capture what people were saying, but to also share these stories with others in the community, city, and beyond. The installation posed one question about culture in the City of Chicago every 15 minutes and asked users to share their responses in the form of stories, opinions, or wishes for the city. The interaction with the installation and

Figure 4.4 The City Listens installation at the Old Town School of Folk Music showing how people engaged with the installation through voice and music.

the recordings of other visitors provided insight on how people view arts, education, and cultural participation in this part of the city.

The City Listens was a two-part system: a primary interface, to be used as a "recording station," and a secondary interface, serving as a "listening station." At the primary interface, participants could use a large touch-screen and a microphone to record their response to a question that was featured prominently on-screen. Participants could choose to share their response in one of three ways: "I remember," "I feel," or "I wish." At any time during recording, participants could pause, stop, or review and submit their responses. Once a response was received, it would be accessible online or through the "listening station," where participants could use an iPad to review, give a thumbs up/thumbs down, or comment on submitted recordings. In addition, responses would become part of an audio loop playing softly in the background at the primary interface, serving to pique the interest of passers-by. Ultimately, these responses were compiled, grouped, and shared with DCASE, where they enriched the larger conversation about Chicago's Cultural Plan of 2012.

Why this installation proved successful was because it created an open-ended loop, offering multiple opportunities to interact with The City Listens both physically as well as online. Participants were not only heard by other visitors to The Old Town School of Folk Music. Their message was carried to a much larger community, as part of an experience that extended beyond the walls of the school, to connect Chicago with culture around the globe.

Both SkyWords and The City Listens were big successes for ID, DCASE, and Commissioner Boone's team. The decision to intervene in cultural hubs around the city meant that many more people were exposed to the cultural plan than traditional events would do. The city was also able to hear from the aggregated voice of minorities, passive cultural participants, and sometimes even reluctant voices. Even more important, the success of the project opened DCASE's eyes to the democratic potential of interaction design and demonstrated how it can be used to empower civic participation in government.

A framework for designing Urban Research Machines

While the projects described above can serve as examples for urban research machines, a key challenge I want to address here is the lack of frameworks that can provide guidance and advice on designing for interactivity in urban contexts. One way to bridge design practice with research is to look beyond technology into user behavior in the wild, to develop insights from observing this behavior, and ultimately utilize the lessons to design future experiences. Through the several installations that we have done over the years, we were able to feed the findings into a comprehensive framework that other designers can use to build such interventions in their own cities. I believe such a framework could potentially help designers in the field to build new and innovative interactive place-making installations.

The following framework is built from the point of view of how the installation once constructed will present itself to the user. The framework is constructed at four different levels that act in progression. At the first level, the installation must entice the user to come to it, and to do so, it must present itself as urban *art,* using material and form configurations that make the installation aesthetically pleasing. Once the user enters into the experience, he/she shifts from peripheral awareness to focal awareness. During this phase, the user expects the installation to present some *information* about itself and its content. Using information, the user can be now persuaded to move from focal awareness to direct interaction and encouraged to engage further. Once the users show a willingness to learn more, the installation must be able to *interact* with the user, and take him/her through the interactive experience housed in the installation. This is where the designer can employ public creativity as a platform for interaction. It must also enable the user to exit gracefully from the experience as well as extend the experience through continuous and extended *engagement.*

This framework is presented here in a diagrammatic manner:

Table 4.1 A framework for the design of Urban Research Machines.

	Art	**Information**	**Interaction**	**Engagement**
What?	At this point, the installation acts as *urban art*; it's aesthetics may come from either the construct of the installation itself or the nature of the interation.	When the user moves into focal awareness, the installation must provide information to answer questions such as "What is this thing? Why is this here?"	Once the user has some information, she wants to learn more. At this point, the installation should be able to support both explicit and personal interactions.	Once people start to interact, the installation should have the ability to engage in deeper dialogue or acts as a catalyst for a dialogue between people.

Why?	Entice people on peripheral awareness thresholds to interact with the installation.	Provide people with information about what they are about to interact with. Help them decide if they want to engage.	Help people overcome social embarrasement, and learn about the urban issue through interaction. Queries can also be posed here.	Enable people to become ambassadors of the issue by bringing others to the engagement.
How?	Installation art, material choice, colour, texture, and light play, innovative interactions.	Explicit messages in the form of written text and images, or cues built into the installation.	Large touch screens, physical interactions, motion, proximity sensing etc.	Synchronous, built into the installation, or asynchronous, so that users can use their own devices to engage.

It is important to note here that the framework is not designed to tell designers what to build; rather, it is designed to work at a conceptual level, to help designers come with conceptual ideas for what an installation might look like. What it actually does or presents is up to the designer.

Conclusion

In this chapter, we started by looking at the changing nature of urban populations both in behavioral and technological positions. Through the lens of two very different but interesting designers, we looked at how public creativity can be used as a catalyst for urban engagement. Using the example of four installations built and installed by the IIT Institute of Design, we explored the idea of Urban Research Machines, and concluded with a framework that can be used by other designers who want to build such machines themselves in different urban contexts.

The value of such Urban Research Machines to civic engagement can be seen clearly through the projects described in this chapter. However, I personally like to look at these installations as baby steps (or prototypes) toward more evolved machines in the future. They do not solve the problem of urban engagement in the modern world, but they move us in the correct direction. The framework should help designers to build better and more interesting iterations of such machines. Whether designers will use frameworks for the design of public creative installations is an important, yet difficult, issue to ascertain. A broad evaluation of any framework will need time and resources that go beyond that which were available for this research. Yet, it is my hope that such urban machines will continue to be developed at the intersection of digital and physical interactions. At IIT Institute of Design, we continue to work with city organizations to explore other versions of such machines; one of our current projects with the tourism department of Chicago will build similar machines at Chicago's O'Hare International Airport to explore traveler attitudes and biases as they fly in and out of the city.

As for Commissioner Boone, the 2012 Cultural Plan was released to the public in fall 2012. It was a great success. Cultural institutions in Chicago are redefining their focus based on the plan, and Chicago is on route to becoming a favored international cultural destination. More important, due to Commissioner Boone and her team's diligent efforts, the process of developing the plan is seen as one of the most inclusive initiatives in Chicago. She may not have reached out to 2.7 million Chicagoans, but more voices were heard and incorporated than ever before in the formation of such plans.

Acknowledgments

The projects described in this chapter would not have been possible if it weren't for the vision and perseverance of several important people in the City of Chicago. Ty Tabing and Lou Raizin from Chicago Loop Alliance were instrumental in recognizing the value of interactive technologies in place making. Jewel Malone, Assistant Commissioner of Culture, and others in DCASE saw the value of connecting interactive place making to civic engagement, and introduced the idea to Commissioner Boone. The framework presented comes from my PhD dissertation at the Open University in the United Kingdom. My supervisors, Yvonne Rogers and Peter Lloyd, receive equal credit for the development of this framework. Of course, last but most important of all are the graduate students who worked on the projects. It would take up much space to thank each of them, but none of this would be possible without their diligence and hard work to design, build, and study these Urban Research Machines.

Image Credits

All images are by the author or project team unless otherwise indicated.

Notes

1 Per the Arts Alliance Illinois, see http://www.artsalliance.org/research/arts-economic-prosperity.
2 Per the WHO Global Health Observatory, see http://www.who.int/gho/urban_health.
3 At www.zerozerochi.com. The website is no longer in service.

References

Anderson, C. (2006). *The long tail: Why the future of business is selling less of more.* New York: Hyperion.

Brignull, H., and Rogers, Y. (2003). "Enticing people to interact with large public displays in public spaces." In *Interact 2003.* Washington, DC: Ios Press.

Chang, C. (2012). *This is the land of Candy Chang.* Available from http://candychang.com/.

Gee, J. P. (2009). "Digital media and learning as an emerging field, part I: How we got here." *International Journal of Learning,* 1(2), pp. 13–23.

Ito, M., Horst, H., Bittanti, M., Boyd, D., Herr-Stephenson, B., Lange, P.G., Pascoe, C., Robinson, L., Baumer, S., and Cody, R. (2008). "Living and learning with new media: Summary of findings from the Digital Youth Project." In *The John D. and Catherine T. MacArthur Foundation Reports on Digital Media and Learning.*

Lozano-Hemmer, R. (2002). "Alien Relationships with Public Space." *TransUrbanism. Amsterdam: NAI Publishers.*

McLuhan, M. (1964/2003). *Understanding media: The extensions of man: critical edition.* Berkeley, CA: Gingko Press.

Rogers, Y. (2006). "Moving on from Weiser's vision of calm computing: Engaging UbiComp experiences." *Lecture Notes in Computer Science* 4206: 404.

Sharoff, R. (2007). "Trying to break the jinx of Chicago's Block 37." *The New York Times,* April 25.

Shirky, C. (2009). *Here comes everybody: The power of organizing without organizations.* London: Penguin Two.

Shirky, C. (2010). *Cognitive surplus: Creativity and generosity in a connected age.* London: Penguin Press.

5
Stories and Conversations in the Smart City

Gerry Derksen, Piotr Michura, and Stan Ruecker

Abstract

If we think of the smart city as a reading environment, we can use it to change what it means to be a citizen, to improve how public topics are addressed, and to democratize how decisions are made. The starting point is text, supplemented with the various other kinds of data that can be gathered through digital means. In this chapter, we discuss two experimental platforms that take different approaches. First is the Data Stories project, where we have been sequencing text from various dynamic sources through a thematic clustering algorithm (Latent Dirichlet Allocation), feeding those thematic clusters into a narrative generator, then putting those results into a storyboarding system. Using the output, we can examine patterns emerging from a variety of text streams, such as Twitter, Facebook, news feeds, and so on. More importantly, however, we can allow people to manipulate the parameters, so that using the same text stream can produce multiple simultaneous valid outputs, depending on the perspective that the reader wishes to take on the feed. Providing a method for encouraging this kind of interpretive or hermeneutic inquiry is a promising strategy for supporting civil discourse. Our second project, Conversational Modeling, is building on previous research to investigate the various ways in which discussions, which occur sequentially through time, can be profitably modeled as 3-D objects of various kinds. These models can subsequently be used for recollection, communication, and analysis, but they may also have a generative potential. As a means of dealing with the structure and substance of discussions in civil society, we propose that conversational modeling has the potential to radically alter our understanding and practice of citizenship.

Introduction

We can read a city, and there are a variety of interesting ways to do it. To take a page out of Derrida's book (ha!), it is possible to consider everything as a text that can be read, and smart cities are no exception. From this perspective of privileging variety, reading can be understood as an ongoing act of interpretation, where the reader brings to the text a theoretical lens that helps to inform understanding. For some people, a city or its specific neighborhoods or streets might be read as an adventure; for others, an investment; for still others, a source of danger. It is not necessary that one interpretation supplant others – in fact, different valid readings from a variety of perspectives are essential to developing a richer understanding. The built environment, the people and their interactions, the public and private elements, the infrastructural systems aspects, and the various forms of commons are all there to be read and interpreted.

However, it is not necessary to go to the extreme of treating everything as text in order to see that the city is an environment for reading. Cities are full of actual text, as anyone can attest who has traveled in a country where they could not read the language. There are obvious and nontrivial city-related texts such as signage for wayfinding, use of public transit; traffic regulation; names of buildings, malls, and stores; public notices; posters, waybills, billboards, and other forms of advertising; menus and receipts; and newspapers. There are also somewhat less immediately obvious but equally important forms of indoor text in cities. These include libraries, bookstores, public records offices, and document archives. Outside, there is informational text associated with public monuments, parks, museums, art galleries, theaters, and other cultural centers of attraction. Finally, and increasingly ubiquitously, there is digital text, whether on a computer screen, mobile device, or public electronic display.

Developing a city story from conversations

There are also the public records of city administration, whether in the form of spreadsheets listing the names of people who are registered as lobbyists, the costs of running the transit system, or the current state of the drinking water. In this chapter, we describe our use of two experimental approaches, not only to reading this kind of material, but also to making it readily accessible and interesting to read, with the ultimate aim of encouraging citizen engagement.

As our case study, we look in particular at the records of three years of discussion of the city council of Charlotte, North Carolina. Charlotte is the largest city in North Carolina, with an immediate population of approximately 800,000, and 2.3 million in its metropolitan region. Home of Billy Graham and an international center of NASCAR racing, it is also the second-largest banking city in the United States, second only to New York.

Our interest in this project was to better understand this unusual place, basing our interpretations on what we could glean using our methods of data+stories and conversational modeling, then considering how these techniques could contribute themselves as inputs to the democratic process.

Process of stories and data

The origin of the data+stories project comes from the idea that memorizing information is made easier from creating stories to help retell the information (Carrell 1984; Mandler and Johnson, 1977; Ross 1986) Many competitors in memorization contests use this technique with playing cards, first to group sets of cards into smaller clusters and then to create a story that helps them remember not only the cards, but also the order in which they appear. For example, if a competitor were dealt a 5 of hearts, 7 of clubs, a Queen of clubs, 2 of diamonds, and a Jack of spades, that person might imagine the following story: Five lovers went to seven different nightclubs in search of dates. They all found the most beautiful to be the Queen of the club who wore two large diamonds around her neck, but she already had a lover named Jack who carried a spade. It is not particularly difficult to remember the five different cards listed here, and perhaps even their sequence, but by creating a number of stories around different sets of five cards, it is possible to recall many more than trying a brute force method to memorize each group of cards and their order. The narrative ties the discrete pieces of data together in a way that is easy to remember and communicate.

We approach building stories by taking data from topics that can be automatically identified in large text collections. The topics drawn from the Charlotte City Council meetings range from public works projects, to civic engagement, to policy making. Any of these topics and more can make up the data points for the story, based on their scope and the amount of coverage they received in the meetings over time. Our data set includes three years (2011–2013). For topic modeling, we used a probability model developed by David Blei (2005), Latent Dirichlet Allocation

Figure 5.1 The final stage in the Data Stories sequence is to produce a visual representation. Here we show the story in the Simulated Environment for Theatre (SET) environment (Roberts-Smith et al. 2013).

(LDA). Having run the data, we selected the top 20 topics, then fed the lists of terms describing these topics into a computational story generator, adopting the point of view of a single character and using parameters from Edgar Alan Poe's story "Descent into the Maelstrom." All topics taken from the LDA process were used in the story as key points of data. From this story we formed an XML-based script, complete with stage directions, to enter into the final phase of the process, where an animatic or moving sketch uses the script to visually represent the story. This phase uses the Simulated Environments for Theatre (SET) software (Figure 5.1). The point of view within the visualization can be changed to illustrate the different perspectives within the story. For example, characters representing a citizen, city council member, environmentalist, or developer can be added to a scene where the discussion concerns a proposed new park system, and the viewer can view the meeting from their visual point of view as they interact with other characters in the scene.

How the story is made

What is a story? Typically, it is a change for a character brought about by experience. In its rudimentary form, a story tells the reader or viewer who did what and where, over time. For automatic story generation, especially involving debate, an important strategy is to include enthymemes, or understandings that are commonly accepted "by and large," in order to lead to a third logical conclusion that is implied. We can also explicitly identify conflicts. Finally, we have the opportunity to use story grammars, or sets of factors that combine to create coherent stories, in much the same way that language grammar helps us to produce coherent sentences.

We do differentiate story from narrative. As Guillemette and Lévesque (2006) explain it, the story generally corresponds to a series of events and actions that are told by someone (the narrator), and represented in some final form, producing a narrative. The actions or environments introduce a conflict that the character must experience and attempt to resolve. Narrative as we define it here is an umbrella term with broader themes that can potentially connect stories. Homer's *Odyssey*, for instance, is a narrative about a journey home, consisting of a series of stories that take place on the intervening islands.

The stories we create serve three purposes. The first is to help the reader understand complex information. A story is created to summarize and present large text data topics as a way to get a grip on the broad complexity of information within the data. The second is to help the reader remember. Third is to introduce the possibility of learning through multiple interpretations, comparing and contrasting stories created from similar data or data flows that are constant.

Using our approach, data topics can be sampled from data flows at regular intervals, allowing us to generate a collection of stories from different time periods. The topics are gathered using Latent Dirichlet Allocation (LDA) which is a three-level hierarchical Bayesian model, in which each item of a collection is modeled as a finite mixture over an underlying set of topics. Each topic is, in turn, modeled as an infinite mixture over an underlying set of topic probabilities. In the context of text modeling, the topic probabilities provide an explicit representation of a document (Blei et al., 2003). From the LDA topics a story is generated computationally using a Markov chain approach (Patil et al., 2010) to randomly select parts of phrases from a given story structure, with the addition of grammar structures. These grammars allow the story to contain a variety of points of view of different characters within the same narrative theme, in order to express the various perspectives and help illuminate the data by suggesting multiple interpretations.

> In multiple narrator texts, conflicts between the reports on the same events by different narrators indicate that at least one of them is unreliable. In realistic literature, a major clash between our world knowledge (extra-textual information) and claims made by the narrator may also serve as such an indicator. (Hansen, 2007: 231)

Often conflicts arise in storytelling from ideologies that contradict each other. The overlap of the narrator's point of view is also often a point of contention, which presses the reader to determine which of the characters is unreliable. Conversely, these conflicts operate as incentives for characters to resolve the conflict for themselves or for others who are trying to identify the focus of the story. Studies in media richness theory show that stories that take advantage of presenting multiple sides of an issue build confidence in the reader and improve performance in understanding (e.g., Dennis and Kinney, 1998).

To take an example of multiple perspectives or interpretations from the Charlotte City Council meetings, a story told by a council member who is a proponent of a project may diminish obstacles that could potentially raise the cost. On the other hand, City Council members opposed to the project may invoke comparison projects to assess cost discrepancies, and citizens who also disapprove of the project may cite environmental or social issues surrounding the project. Further still, business leaders might counter with economic advantages and secondary revenue generated by the project.

The environment is also an important factor in a story. It can encourage or hinder obtaining a goal, but the salient characteristics can change based on which character is selecting them. Between characters, conflict, and environment, a story is generated to depict the behavior and interaction between the stakeholders, problems, and contexts suggested by the topics generated from the minutes. We can see the correlations by comparing the minute transcripts and the stories generated from the topic descriptions. As we subtly refine the story generator, shifting between the points of view and character description, we generate a richer summary of the document data.

LDA topic distributions from Charlotte City Council meetings 2013

topic #1 foundation, committee, board, best, groups, past, fact, last, need, community

topic #2 wideman, terminal, management, buses, revised, 19:, plan, environmental, video
topic #3 fallon, ms., sir, engineering, main, agreement, acclamation, range, appoint, administration
topic #4 defer, recuse, right, 20, voting, absent, noted, you, Howard, contracts
topic #5 councilmember, motion, seconded, made, Barnes, Kinsey, cannon, Howard, Mitchell, Mayfield,

topic #6 add, Natalie, can., energy, staff., duke, 2013., active, e., 28:
topic #7 15, June, 30, beginning, live, rail, light, houses, immediately
topic #8 2011, planning, charlotte-mecklenburg, ordinance., department, adopt, submitted, marketing
topic #9 July, llc, water, fee, storm, rental, adjournment, court, development,
topic #10 roofing, forum, reconvened, parking, decisions, citizens, Andy, Michael, looks, Lawana

topic #11 tax, said, mr., hard, guys, wanted, bring, really, ms., say
topic #12 8, okay, mixed, 5, hope, check, start, partners, said, request
topic #13 numbers, takes, helped, yet, taxes, behalf, publicly, design, now, honored
topic #14 sc, capital, tem, meetings, question, amending, clarify, wait, program, agencies
topic #15 councilmembers, Kinsey, Mitchell, nominated, Dulin, Cooksey, Autry, Pickering, cannon, Barnes,

topic #16 ordinance, recorded, follows, full, book, page, ballot, cdbg, funds
topic #17 value, parking, there, mr., said, city, approved, asking, development
topic #18 project, project greater, putting, appointed, policies, places, stop, net
topic #19 bids, summary, construction, Walton, united, inc., curt, manager, bullseye, city
topic #20 modified, Mcalpine, recommended, ability, mallard, obligation, material, reporting, lighting

Many stories for many citizens

What is the purpose of creating stories of this kind? According to Sitkin et al. (1992), stories increase the capacity for carrying data. They also possess symbol-carrying capacity, providing information about the information or about the individuals who are communicating (Sinclair, 2005). If the primary goal of a story is to assist memory, it is less important that the story is believable or even logical. However, even a fantasy holds specific pieces of information that stay intact. According to Hansen, Norlyk, and Wolff Lundholt (2013), in a communicative context, storytelling enables organizations to establish dialogical relationships with multiple stakeholders. That is, the overall strategic purpose is to use and control stories inside and outside the organization in order to establish long-lasting, value-based relationships with different groups of stakeholders in order to strengthen the corporate brand and differentiate the organization from its competitors. In a similar way, we see cities attempting to control the narrative to present initiatives, projects, and trajectories to its citizenry. As cities tell stories, consensus can be built among different user groups around the validity of the data present in the story. Data points within stories add to the believability and conversely the story surrounding the data points helps people to remember the data.

> Narratives are useful data [in themselves] because individuals often make sense of the world and their place in it through narrative form. (Bruner 1990; Gee 1986; Mishler 1986; Riessman 1993.) Through telling their stories, people distill and reflect a particular understanding of social and political relations. (Feldman, Skoldberg, Brown, and Homer, 2004: 148)

Where stories often become disputed is in the surrounding information or discourse structure that binds data together within the story schema (Brewer and Lichtenstein, 1980). According to this line of reasoning, discourse structures are more than just supporting words to provide the chronology of events (event structures). Instead, they form the framework of what we recognize as a story schema. Narratologists use the Russian Formalist term "fabula" to distinguish story schema from "sjuzhet," which is the representation of events within the fabula. The fabula of the story supports the rationale, assumptions, emphasis, and values of the storyteller. If city stakeholders' acceptance of the data builds consensus in the discourse structure of city council meetings, challenges of the framed problem will be found in the story fabula. We see this in the minutes of the meeting transcripts where members respond to objections (whether real or imagined, past or predicted) as a way of strengthening their position in support of the narrative they are building.

> I've even heard the question one time, "Well, Fulton, if you do mixed-income housing the outside of the building will look the same, but

> inside will look different." That is absolutely not correct. The inside and the outside of the building will be market rate; it just has an affordability metric built inside of it. (Feldman, Skoldberg, Brown, and Homer, 2004: 148)

In addition to the data points themselves, the relationship between the data points provides a point of reference that can help to legitimate the story. This is especially important where the story condenses the broad view of council meetings into a concise story. In Fulton's statement above, the highlighted data points are mixed income, market rate, and an affordability metric. These are framed opposite the challenging question at the beginning of the statement. "...if you do mixed-income housing the outside of the building will look the same, but inside will look different." The question is raised as a straw man that can be dismantled by the emphatic "absolutely not correct." In contrast, the terms of the data points are highlighted because of their business-like tone and what appears to be a sympathetic view toward people of lower incomes.

Just as there are methods to create suspenseful, surprising, exciting, or melancholy stories, by ordering information we use methods of building a story to argue a point of view. One such method is from classical rhetoric, which uses enthymemes as a basis for the story. Enthymemes are defined as a kind of loose logic that uses two points of "common knowledge" to lead to a third implicit statement. To change the philosophical lens of the story, we can include enthymemes and still maintain the data's validity, while at the same time questioning the logic because it is not necessarily accepted let alone "common" among outside groups. For example, in the following story developed from the Charlotte City Council meetings, topics stay consistent, while the story structure or fabula used to embellish the narrative is manufactured to support a heroic adventure around bus transit.

Computationally generated story

> We built the foundation around the committee board, we're now in the teeth of the best group, whose past, as far as the facts of my invariable miscalculation, would set the Mayor upon a needed calm, which is perhaps among the most amazing of Charlotte communities. Twice, during 19 years, the plan was completely revised, and all this time I held my breath, and clung to management, "for I have brought you to the terminal that you might send buses out into the environment." They say too that the other engineers had entered

> the whirl of the buses, voting within their right for contracts which recused me at first. Immediately our light rail gave herself a shake, just as death-condemned felons in prison are allowed petty indulgences, administration appointed council members while their doom is yet uncertain. To the right and left, as far as the vvurrgh of buses lashed out into ungovernable fury; but it was in me to triumph – which was undoubtedly the planning ordinance of my department which confounds Mecklenburg County. The oldest council members in Charlotte never experienced anything like it. All this time I had made sure that the Mayor was on board – but now we were planning for parking, decisions which we submitted to the council.

Ms. Natalie was quite upon the transit storm; busily amending questions or clarifying these observations, rendering me anxious to turn them toward agencies. For every revolution, we passed something like a cannon, or a full book or the pages of a vast ballot of votes to approve a tax – no, the oldest council member in the city of Charlotte never experienced anything like it. In fact, we made it a matter too difficult to stop, the anecdotes of the project purveyors, which could not have approved the bids of my policies, that I recommended the construction of the material by Mcalpine, now lay flat and frothing, only to get up to meet my manager obligations.

Topic terms from LDA and sentence fabula from E. A. Poe's "A Descent into the Maelstrom"

Despite the flowery nature of the text, there are a number of sentences that suggest ideas that are there to establish the questionable contrast or validity of the hero's behavior, "...would set the Mayor upon a needed calm." This sentence suggests that if the transit issues were solved the Mayor could relax; however, few would assert the position of city mayor can be described as calm. Later the sentence, "...the oldest council member in the city of Charlotte never experienced anything like it," proposes the protagonist is unique in this approach to handling the bus system. The statement may be true, but it is really in contrast to the age of the council members that suggests the protagonist is progressive in this new way of thinking. "I recommended the construction of the material by Mcalpine" informs us that the recommendation is justified because it was done to ease the job of the mayor and is forward thinking compared to the rest of city council. To be sure, those who tried to stop our hero were thwarted: they "now lay flat and frothing." If we accept the position that the hero is uniquely approaching the problem and that newer (younger) equates to better, then we should follow the "logic" and award Mcalpine the contract. The enthymeme and the counterposition would suggest, however, that new and unique solutions may in fact be just naive.

When considering any number of views surrounding city issues, it is useful to weigh the validity, viability, and passion of often outspoken perspectives voiced by citizens. Stories backed by data have two benefits when considering the citizen view. First, the stories generated can be drawn from differing points of view to be inclusive and attempt to incorporate different voices – especially those in opposition. There is the protagonist view, which, typical of many stories, may be heroic, metamorphosing, or humbling, and there is the antagonist view which can often be pedantic, demoralizing, or destructive. The ability of stories to change the way we perceive information is supported in Denning's argument that organizational change is often based on taking alternative perspectives: "transformation requires organizations not just to learn but also to unlearn, to rethink how and even why they undertake certain activities.... [W]e need to unlearn practices and mental frames that we don't even realize we rely on but which shape our whole perspective" (Denning, 2000: 8). In this context, Sole and Wilson (2002) identify the following five values that stories convey in terms of both information and emotion: asserting norms and values, developing trust and commitment, sharing tacit knowledge, facilitating unlearning, and generating emotional connection.

The second benefit to data-anchored stories is the consensus built around the data points. From our story example, a number of statements may be disputed based on how they are handled by the storyteller, but the basic tenets of the story still hold true – the transit system contracts were negotiated by the city and a new light rail was part of the discussion. How these points are described as positive or negative is part of the debate, but for some, we feel support for the triumph in the "vvurrgh of buses."

Advantages and disadvantages of data stories

Generating stories to inform and persuade people is not a new concept. However, we address some of the shortcomings of computational narrative generation by introducing the LDA topics as points of reliable data. The advantage of computational story generation is the ease with which multiple views can be generated and a kind of serialization of stories can be consistently delivered. Supporting memory through mnemonics and taking advantage of larger data capacity are among the short-term benefits of storytelling. Analysis of multiple stories, either from various stakeholders or from one position over time, we see the potential for building confidence in the civic process, opening dialogue between parties, and forming consensus between city officials and citizens.

Taken together, multiple stories show a city's much longer story arc to compare changes in data (topics) and changes in plot that slant toward one character's view over another. This requires a more refined story generator that formulates a more temporal structure of narratives around the city's many stories. The temporal aspect of narrative that follows real-world time is described by Ghislotti (2009) this way: "the episodic memory system of encoding strictly keeps the temporal connections between events, and only a willing mental operation can dispose them in a different order. Using this memory function, we are able to manage a mental representation of the fabula." In our discussion, we have contrasted the story and data to emphasize the points in the story worth remembering and to characterize the narrator's point of view. As we work toward a more sophisticated story generator, the fabula, sjuzhet, and schema will align more closely to represent the story told by the people of Charlotte, but the value of understanding how multiple stories change over time is arguably at least as important as any individual story.

Conversational models from citizens and city leaders

Where the data+stories project attempts to produce interesting and memorable summaries of material that would otherwise be prohibitively long and dull, the goal of conversational modeling is to dive deep into specific moments of the prolonged discussion. The process focuses on the details of conversations, and in several respects, city governance can be understood to proceed by conversation. In some instances, these conversations take place behind closed doors, between elected officials, lobbyists, policy makers, or other administrative personnel. In other cases, the conversations are held in public and can involve interlocutors who may on the one hand be members of the official leadership, or else on the other hand may be interested citizens without a leadership mandate.

In principle, policy formulation should involve a dynamic conversation between all interested stakeholders. It is the latter form of conversation that we modeled, using a physical modeling technique, which allows the people doing the modeling to interpret the concepts that frame a discussed topic as presented by conversants, and then broaden the scope of possible outcomes of the conversation by questioning assumptions and articulating other perspectives on the issues at hand.

Gordon Pask (1976a), a cybernetician and a proponent of conversation theory, understood conversation as an exchange of points of view showing and clarifying differences between autonomous individual participants. Building distinctions among participants is based on one participant's understanding of what the other participant has in mind. When this important stage of conversation is achieved (namely recognition of differences) the conversation already is successful even if both parties just "agree to disagree." Pask (1976b: 23–24) pointed out that participants' conceptual structures, which are basically what both parties know, are inaccessible and only their representations can allow direct examination. The conversation can be seen as a constructive process due to the fact the understanding can be built only according to interpretation of symbolic representations provided by the participants. Between participants of a conversation there is a "conversational domain" (Pask 1976a: 19), which is a space in which conversation takes place. Spoken language creates this kind of space, but so can a mediating system with a visual interface – as in the case of our model. It is pointed out by Pask that language used and understanding cannot be taken for granted and usually are negotiated by participants in the conversation, which has its own dynamics (Richards 2007: 133).

Ideally, for a good conversation to take place, there must be some points of overlap in participants' understanding of some concepts, using shared language. One of the main issues is a language used for interaction, which would be able to represent participants' thinking about the issues at hand, and allow comparison of positions. For Pask (1976b), the Repertory Grid technique developed by Kelly (1955/1991) was one of the possibilities to elicit personal constructs about a particular topic, represented in the participants' own terms.

The Repertory Grid was devised by Kelly and is based on his Psychology of Personal Constructs, which was an attempt to figure out how people construe their understanding of events. The main postulate presented in the theory of personal constructs was the notion that "a person's psychological processes are channelized by the ways in which he anticipates events" (Kelly 1955/1991: 32). It stresses the individual point of view in acquisition of knowledge – construing reality. People develop their own theories about their surrounding world, test those theories, and update them according to their experience. On the basis of those theories they anticipate events and act accordingly. The theory is a constructivist approach because it is focused on personal views on events, envisions people as active agents making sense of events around them, and is applicable for a particular purpose or circumstances (Butt 2008: 14). The Repertory Grid was one of the tools developed in order to operationalize the theory. The grid can document internal cross-references between a set of personal constructs. Shaw (1980) points also to potential use of the grid as a conversation support tool. "Used in conversational mode the grid can be an articulator of conversation, the clustering of responses providing a starting point for discussing individual differences and points of view" (Shaw 1980: 23–24).

The method for eliciting how people construe topics is based on identification of difference poles. The poles are placed on opposite ends of scales, so each scale consists of contrasting opposite descriptors, which are meaningful for the participant in relation to the topic. The scales are then used to score/evaluate particular elements chosen by a user, which instantiate the topic as seen by the user.

We used the Repertory Grid technique to analyze conversations in a nontraditional way. We had at our disposal the large set of transcribed dialogues from the actual conversations taking place during the city council meetings. Following the repertory grid procedure in the reverse order we tried to interpret the city council members' constructs out of their speeches. The question was what kind of concepts are behind what was said in a particular moment of a conversation. We aimed at figuring out patterns in their way of discussing things to ask questions about the factors that could shape such patterns.

In the case of our model, the poles were determined from comments that summarized the topics within the conversation. Following Kelly's method, we continued to set differentials until all of the topics were satisfactorily covered. Then, we modeled the conversation from all contributing speakers, treating individual comments as text "snippets," and placing them on respective scales in places, where intensity and direction toward a particular descriptor in the scale indicated interpretation of text meaning.

The modeling can inform the observer of the range in topic domains that the conversation naturally takes, as well as reveals patterns and connections not easily made by reading the transcripts. Setting up scales with descriptors on opposite ends form boundaries of conceptual space in which the issues are discussed.

The approach described above aligns with Richards's (2007) account of how design of social systems can be understood. Richards adds to the understanding of the policy formulation process the notion of constraints, first as general understanding of policy as agreed constraints on behavior, second as a viable method of dealing with complexity and variety of societal issues. Policy formulation is a process of setting up constraints on decision space. Richards (2007) reconsiders the prevailing model of policy formulation, which puts stress on goals and outcomes (a goal-oriented approach) and instead he proposes an approach based on constraints creation and analysis (a constraint-oriented approach).

The latter is conceived in the tradition of cybernetics. Richards traces this idea back to W. Ross Ashby and his concepts of variety and constraint. Information was conceptualized as reduction in variety by constraining the possible available outcomes. Ashby observed that often analysis of complex systems is performed more efficiently by identifying constraints rather than identifying all possible relations within the system. This has been called negative thinking – the observer notes what is actually excluded within the system and focuses on what is not wanted or is undesirable.

In terms of policy formulation, this approach allows us to question and reconsider the boundaries in which a particular policy is formed, to identify hidden assumptions, and to look at possible outcomes from different perspectives, possibly involving other stakeholders and address their concerns.

It is important to clarify that these ideas are rooted in a radical constructivist approach, which states that the only possible understanding of the world people live in is accessible through their experience. To this way of thinking, direct access to objective reality is impossible because there is nothing like an "objective reality." There are only constructs. People construct their understanding based on what is available to them at the moment and what works for them in particular activities. Knowledge they possess is in fact an adaptive instrument – it helps them to operate in the world, and constructs persist which are useful, viable. Additionally, those concepts, which they form about the world, come from embodied experience, constructed on the basis of acting in the world. So they are to a greater or lesser extent different among people. Each person construes a mental image of the world, which in turn regulates perception, by how it allows them to anticipate events, and thus constrain and guide subsequent behavior.

Instead of shaping policies in a paradigm of sequential backward thinking, by starting from values and desires reified in goals and objectives, Richards proposes the approach that puts values as constraints in a model and allows stakeholders to reconsider their implementation in many alternative ways. The method enables "a presentation of the range of possible outcomes or behaviors that could be accommodated within a set of constraints and that could change or be changed as a consequence of an action" (Richards 2007: 131).

The constraints shape can be thought of as a decision space. It means that possible policies can be formulated within the boundaries of the system set by an observer by his/her decisions about the set of constraints.

The model allows broadening the scope of the possible outcomes by direct inclusion of the opposite poles implicit in constraints (i.e., looking at "what is not wanted or undesired") and then particular scenarios can be imagined in which the previously "unwanted" can be actually "desirable" (Fischer and Richards 2013: 6).

This is exactly how we consider the use of the Repertory Grid technique in our conversational model. The way concepts are considered by conversation participants, the perspectives, may differ also in dimensionalities involved in the space of understanding. The Repertory Grid technique, by setting difference poles, is aimed at revealing those dimensionalities of personal constructs about issues under discussion, as well as the specific language used for expressing these differences.

Referencing the discourse in detail – conversational modeling

In particular, we focused on a conversation about Charlotte's bicycle-sharing program that occurred over the course of a year. Like other conversations, whether in business, health care, law, education, and so on, these can be considered high-value conversations, a kind of enterprise asset that has not necessarily been given the attention it warrants. At the moment, the best practices for capturing and communicating conversations are largely sequential, whether in the form of video, audio, or transcript.

There are also, however, a variety of nonlinear methods that have been the subject of experiment and use over the years. These include various forms of mapping, often described as mental or cognitive, as well as more general types of diagramming, such as the somewhat novel approach known as graphic facilitation, where an artist will attend a meeting and produce an illustrative summary that may feature scenes, characters, speech or thought balloons, and so on (e.g., http://drawingoutideas.ca/services/). One drawback of these approaches is that the modeling is not always given close attention.

The city of Charlotte adopted a shared bicycle program in 2011, coordinated by the B-cycle organization, with city support but funded by private donations. A conversation between six city councilors, the mayor, and the director of B-cycle who reported on the progress and future direction of the program was captured in the city council minutes (Figure 5.2).

Following the Repertory Grid method, we set differentials until all of the topics were satisfactorily covered; however, we organized these

Figure 5.2 Charlotte B-cycle station which includes stalls for 20 bicycles, map of the city with other stations, and payment kiosk.

topics into high- and low-level categories to consider the priority of a comment relative to previous comments in the discussion. In earlier conversational modeling studies we have seen that referencing authoritative sources or members of the group happen regularly to establish current beliefs or assumptions. (e.g., Derksen et al. 2013) As a result, we can see from the model where high-level comments connected to lower levels, as well as where comments were clustered around central ideas.

As mentioned before, we modeled the conversation from all contributing speakers, treating individual comments as text "snippets." These snippets are color-coded to indicate the speaker, and are placed on a continuum between subtopics within broader subjects of money, infrastructure, and program use. The subtopics were then divided into ideas determined to be at opposing ends of the continuum. For example, "expansion plans" was at one end and "underperforming stations" was on the other. Six total subtopics were generated based on the occurrences of snippets included in the minutes under the heading of money: privately funded, pricing structure, nonpayment policy, annual membership, leasing agreement, and student rates. Subtopics in the infrastructure group were 20 stations, 200 bikes, promotions, expansion plans, statement of principles, and underperforming stations. Finally, the use topic was made up of rider feedback, new membership, member use, and optimal use strategy (see Figure 5.3). Once the subtopics and their opposing terms were structured on a three-dimensional grid, each snippet was placed on the model between differentials made from the subtopics.

Figure 5.3 Speakers who participated in the conversation represented in the model; Director of B-cycle – Ms. Ward, Council members – Mr. Dulin, Mr. Autry, Mr. Cannon, Mr. Pickering, Mr. Reiger, Mr. Howard, Ms. Kinsey, and Mayor Pro Tem Mr. Barnes.

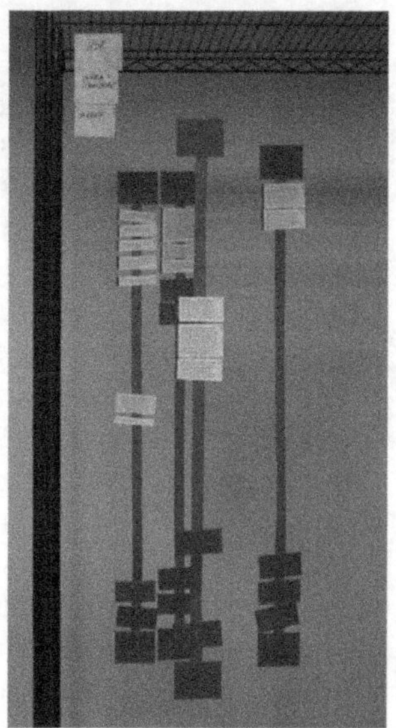

Figure 5.4 A detail of user input positioned on scales of opposite constraints to frame the topics within the model.

In Figure 5.3 we see labeled connectors showing relationships between the subtopics under the main topic headings. In this example, B-cycle has more connections between student discounted prices enticing potential riders to increase ridership – emphasizing use over profit. Encouraging ridership indicates a subtext of Ms. Ward the presenter of the program: "we're evaluating each one of the bottom ten performing stations and identifying is it something that will be corrected as we expand or is this something that won't have a solution at all." As her comment suggests, there is some concern that the program is still in the early stages of growth and the city is unsure of the reason behind underperforming stations. The presentation, however, balances promoting ridership with a view that is optimistic and often celebratory. Uses of the phrase "we want to celebrate that" when describing positive data reports indicate Ms. Ward's eagerness to deliver the information in a positive manner which is often reciprocated in congratulations from the council members and the mayor.

A strategy for setting up the share system also hints at promoting use over profit: "So we encourage all people, both guests and Charlotteans to make sure that they dock the bikes every 30 minutes to avoid those charges." If riders who have a membership dock in between rides they do not get charged for the first half hour. Placing the stations 30 minutes apart at an easy pace encourages this behavior while saving their users money.

Some questions concerning riders not paying, underperforming stations, and students not able to afford the fees come up during the conversation. "Mr. Dulin said...I saw the bikes up at JC Smith the other day and I would just sort of think that it would be cost prohibitive for a college student." Ward uses these comments to segue into data she had already planned to present, "I am glad you brought up the issue of student fees..." and continues on to make her point supporting the students as a good way to introduce "Charlotteans" to the service, "They are eating Ramen noodles, they are not going to try to pay $4, that $4 is a meal. So, our goal is to get students to utilize the system because one of the things we know is that students will ride bikes."

The presented coverage of hierarchy of topics and subtopics is then mapped in the space constructed by identified dimensionalities of constraints (Figure 5.4). As city officials report their achievements, it is possible for the observer/user of a model to identify values they are following in setting up the constraints on the space of possible outcomes. It is also possible to ask about the opposite poles, which are inherently implicit. So if one of the values, which are unquestionably assumed, is the affordability and accessibility of the B-Cycle system, we can ask if there are any possible scenarios, which could make acceptable that the system is expensive and elitist. What could be the advantages to the city, and what kind of users may appreciate this idea? Following this way of thinking, we may consider that perhaps a more expensive system can provide a better service, for example, more stations and bikes can actually be implemented. Of course the model is not meant to promote totally opposite solutions, but just to expand the number of possible alternative. The success of its use would be satisfactory even if a slight change in policy can be achieved which at least partially addresses a hitherto silent group of stakeholders.

The presented physical conversational model visualizes the space of possible outcomes, allowing the user to consider many more options than have been presented by the city officials and at least question some of the decisions already made in order to further upgrade the B-Cycle system.

Shaping the city through text analysis

The goal of our approach is to combine a high-level strategy for identifying, interrogating, and remembering key topics (i.e., data+stories) with micro-views of specifically selected topics, where the details of the interactions between stakeholders are brought into a structured conversational model. Ideally, this combination of overview and detail can be leveraged over time as a means of helping to direct future decisions.

Attempts of this kind are increasingly useful as data from cities proliferate. A number of data collection systems have cropped up in cities around the world: kiosks, cameras, surveys, websites, apps, and sensors all collecting and collating information gathered from citizens. Much of the data capture behavior, which says a great deal about the actions of the people involved in a given activity. What is often unseen is the perspective, motivation, or information the actors are operating from. We would assert that without the textual input of the people, it becomes difficult to assess where the behaviors originate and what communication will change the stories they have constructed. The creation of data stories and modeling conversation is an attempt to construct a mental space around the subjects that are important to all the stakeholders. Within these spaces, we see how the emotional tenor of the conversational model is revealed. The meetings are often charged with exuberance about a project completed successfully or the tension of making difficult policy changes.

What is important within story generation is that we see them as an opportunity for shaping the general direction of a city, particularly if we can compare stories over time. The conversation models indicate how the city engages with the participants as individuals; if plotted over time, these should correspond to the stories' progression. In the next phase of our work, we plan to follow some of these trajectories to determine the extent to which the data stories can align with the conversational models.

What is essential for cities, whether through the techniques we propose or others, is to "read" the values of their citizens and find ways both to collect information and to respond to their emotional characterizations. Collectively, these conversations begin to build a story shared by communities that make up the city. If these stories are in stark contrast to the stories the city is trying to construct, we see how well-intended initiatives fail. If communities desire art spaces, local café owners, and unique artisan shops but the city sees potential for art dealers, craft fairs, and larger commercial events for a design industry, it is possible to see either how these two stories could align or else how they seem out of sync. A city can benefit from seeing how individual contacts with citizens have contributed to the overall direction it has taken. Similarly, if the city can proactively see potential for more citizen engagement in the direction it is headed, it is to everyone's benefit. If citizens are empowered to influence the decisions made by the council members, then a shift toward more supportive communities can occur, rather than a dissatisfied electorate who votes members out of office. Finally, groups in the community that are currently isolated may start to speak with a stronger voice if more supportive voices could be identified.

Conclusion

The two methods presented in this chapter are a micro- and macro-view of a city through text analysis, which we propose could result in improvements for democracy. Specifically, for our case study, we looked at points of view in city decision making. From the macro-level of multiple stories told from various points of view we begin to make visible the direction a city maps out for itself. We believe that people within the city will adopt stories of this kind, if they think the stories represent the values they too hold for the city, and if they can contribute in some way. Similar shifts toward democratization of providing data to online community stories may encourage contributions and continual rewriting of city stories. Access to city leaders and feelings of being heard can be strengthened using online town hall meetings as opposed to attending city hall meetings. The method proposed does not advantage one particular view but will only be generated given a prevalent and persistent nature of a topic on the city council agenda.

In contrast, the micro view of conversational modeling privileges the individual speakers within a topic and balances the tenor of the generated stories with actual accounts of conversations. Inclusion of the emotional evaluation of conversational modeling adds the dimension of depth to the sentiments held by citizens or city councilors not easily seen in the generated story. Conversational modeling, like story generation, can indicate arcs of a story; however, the conversational model emphasizes the immediate focus of the issue. As more conversations about a subject arise in city council meetings, people studying the conversational models will begin to see how they affect the overall telling of the city story. Once made visible, any proactive and prolonged attempts to propel a single agenda may result in decisions to more fully inform all sides of new initiatives. Such strategies would give citizens more opportunity to counter or support such initiatives, which is a positive outcome for a more democratic approach to making decisions about city planning. At this point, the technologies are in their infancy; as they mature, their potential to play a role in the lives of cities, and citizens, will grow.

References

Blei, D., Ng, A. Y., and Jordan, M. I. (2003). "Latent dirichlet allocation." *Journal of Machine Learning Research* 3: 993–1022.

Brewer, W. F., and Lichtenstein, E. H. (1980). "Event schemas, story schemas and story grammars." Center for the Study of Reading, Report 197. Urbana, IL: University of Illinois at Urbana-Champaign.

Butt, T. (2008). *George Kelly. The psychology of personal constructs*. New York: Palgrave Macmillan.

Carrell, P. (1984). "The effects of rhetorical organization on ESL readers." *TESOL QuarterlyI,* 18(3): 441–469.

Denning, S. (2000). *The springboard: How storytelling ignites action in knowledge-era organizations*. Oxford: Butterworth Heinemann.

Dennis, A.R. & Kinney, S.T. (1998) Testing Media Richness Theory in the New Media: The Effects of Cues, Feedback, and Task Equivocality. Information Systems Research. Vol. 9, No.3, September.

Derksen, G., Michura, P., Ruecker, S., Pollari, T., Dobson, T., Musick, H., and Audette, S. (2013). "The role of conversational sculpture in design education." *Proceedings of the International Association of Societies of Design Research* (IASDR), Tokyo, August 26–30.

Feldman, M. S., Skoldberg, K., Brown, R. N., and Homer, D. (2004). "Making sense of stories: A rhetorical approach to narrative analysis." *Journal of Public Administration Research and Theory*, 14(2), pp. 147–170.

Fischer, T., and Richards, L. D. (2013). "From constrained design to designed constraints to constraints reversal." *Proceedings of the DesignEd Asia Conference*, Hong Kong Polytechnic University, December 3–4.

Ghislotti, S. (2009). "Narrative comprehension made difficult: Film form and mnemonic devices in memento." In Warren Buckland (ed.), *Puzzle films: Complex storytelling in contemporary cinema* (pp. 87–106). Oxford: Wiley-Blackwell.

Gonzales, D. (2000). "Story grammars and oral fluency." *The Journal of Imagination in Language, Learning and Teaching,* 5, pp. 74–79.

Guillemette, L., and Lévesque, C. (2006). "Narratology." In Louis Hébert (dir.), *Signo* (Rimouski,Quebec, Canada). Retrieved from http://www.signosemio.com/genette/narratology.asp.

Hansen, P. K. (2007). "Recognising the unreliable narrator." *Semiotica* 165, pp. 227–246.

Hansen, P. K., Norlyk, B., and Wolff Lundholt, M. (2013). "Corporate storytelling." In Peter Hühn et al. (eds.), *The living handbook of narratology* (paragraph 5). Hamburg, Germany: Hamburg University.

Kelly, G. A. (1955/1991). *The psychology of personal constructs,* Vol. 1. London: Routledge.

Mandler, J., and Johnson, N. (1977). "Remembrance of things parsed: Story structure and recall." *Cognitive Psychology* 9, pp. 111–151.

Margolin, U.. "Narrator." In Peter Hühn et al. (eds.), *The living handbook of narratology*. Hamburg, Germany: Hamburg University. Retrieved from http://www.lhn.uni-hamburg.de/article/narrator.

Pask, G. (1976a). "Conversational techniques in the study and practice of education." *Journal of Educational Psychology*, 46, pp. 12–25.

Pask, G. (1976b). "Conversation theory." Amsterdam: Elsevier.

Patil, A., Huard, D., and Fonnesbeck, C. J. (2010). "PyMC: Bayesian stochastic modelling in Python." *Journal of Statistical Software*, 35, pp. 1–81.

Peinado, F., and Gervás, P. (2006). "Evaluation of automatic generation of basic stories." *New Generation Computing*, 24(3), pp. 289–302.

Poe, E. A. (1910). "A descent into the maelstrom." In *The works of Edgar Allan Poe*. New York: Harper and Bros.

Richards, L. D. (2007). "Connecting radical constructivism to social transformation and design." *Constructivist Foundations*, 2(2–3), pp. 129–135.

Roberts-Smith, J., DeSouza-Coelho, S., Dobson, T., Gabriele, S., Rodriguez-Arenas, O., Ruecker, S., Sinclar, S., Akong, A., Bouchard, M., Jakacki, D., Lam, D., and Northam, L. (2013). "Visualizing theatrical text: From watching the script to the Simulated Environment for Theatre (SET)." *Digital Humanities Quarterly*, 7(3). Retrieved from http://www.digitalhumanities.org/dhq/vol/7/3/000166/000166.html.

Ross, S. (1986). "An experiment with a narrative discourse test." Language Testing Research Colloquium, Institute of Education Science (ERIC), Monterey, CA.

Shaw, M. L. G. (1980). *On becoming a personal scientist*. London: Academic Press.

Sinclair, J. (2005). "The impact of stories." Swedish School of Economics and Business Administration, Helsinki, Finland, *Electronic Journal of Knowledge Management*, 3(1), pp. 53–64.

Sitkin, S., Sutcliffe, K., and Barrios-Choplin, J. (1992). "A dual-capacity model of communication media choice in organizations." *Human Communication Research*, 18(4), pp. 563–598.

Sole, D., and Wilson, D. G. (2002). "Storytelling in organizations: The power and traps of using stories to share knowledge in organizations." LILA, Graduate School of Education, Harvard University, Cambridge, MA.

6

Are Creative and Green Cities Also Smart and Sustainable?

Kevin Stolarick and Olga Smirnova[1]

Introduction

Clean, green, creative, and now smart cities have all been separately identified, measured, ranked, and evaluated. Are they really all just different ways of talking about the same things? Cities that rank well in one category always seem to do well on the others. This research identifies and compares creative, green, and smart cities and looks for correlations. This research will proceed along two lines. First, new measures for identifying smart cities and sustainable cities are developed. These are then combined with existing measures for green and creative cities. Which cities are creative but not smart? Green but not creative? Second, what are the relationships among cities for being creative, green, and smart? This research will help cities, regions, and policy makers, many of whom are pursuing growth strategies based around one or more of these concepts. By discussing the relationships among these strategies, a more nuanced approach may be developed.

According to Frost and Sullivan[2] top 20 mega trends, the green concepts will be replaced by "smart" in the future. However, without having the ability to measure either concept empirically, it is not clear whether the words will substitute each other versus the actual emergence of a totally new urban phenomenon. Thus, we need to find ways to measure all concepts before we will be able to determine their impacts. This chapter proposes innovative ways to measure both green and smart cities concepts. Our findings contradict the modern forecasts: smart and green are not the same, so they cannot be necessarily replaced by each other.

Smart cities

A smart city is a place where "ICT [information and communication technologies] is merged with traditional infrastructure, coordinated and integrated using new digital technologies"[3] (Batty 2012, p. 1). The fast development of ICT[4] is transforming our daily lives and our communities by minimizing transportation and communication costs. There is a number of different initiatives by private companies (e.g., IBM Smarter Cities[5]), universities (e.g., MIT cities initiative[6]), and various governments (e.g., European Union Innovation Partnership on Smart Cities and Communities[7]) devoted to both taking advantage of rapidly developing ICT to make cities smarter and to understand the changes that new transition brings. "We use technologies to study cities and develop new products, the cities and human interactions ultimately change and we still do not understand neither the change nor our role in it" (Batty 2012, p. 36).[8] This is the space–time convergence (Batty 2012, p. 18) that makes the science of cities so difficult and so fascinating. With technology costs being reduced significantly within the next 50 years, and population and energy demand are projected to increase significantly within the same time frame (UNCSD 2011), smart cities become a promise of the new technology to solve the problems with increased resource constraints.

The promise of deploying ICT to help solve pressing urban problems leads various organizations to combine different characteristics that will be shared by smart cities. For example, Smart Cities Council states that future cities will have the traits of "livability, workability, and sustainability."[9] Federal Highway Administration Livability Initiative[10] (FHWA) ties the quality and location of transportation infrastructure to access to quality jobs, affordable housing, and quality schools. The livability refers to the physical amenities that lead to a better quality of life. Michael Batty (2012, p. 12) calls quality of life the "mandate" of the cities. The quality of life is improved in the smart cities through increased mobility and access (Batty 2012). "Workability" is related to connectivity, access, transportation, and technology, while "sustainability" refers to "low-impact living" and efficient use of resources. In other words, the smart city of the future is a clean place with convenient 24-7 digital services, and transportation and communication technologies enhancing the quality of life in the place. One of the ways this can be implemented is through the bottom-up approach (Batty 2012) or dynamic incentives for shared use of resources (MIT). The European Union (EU) Innovation Partnership for Smart Cities stresses energy efficiency, better planning, better transportation, intelligent use of information technologies, applied innovations, and social networks as the result of ICT implementations at the city level.

How would smart cities achieve all this good life for everybody? It would be through the interaction of four important mechanisms discussed further below.

Innovation and technology

One of the first components of smart cities is technology that changes the way services can be delivered and monitored 24-7. The modern ICT allows big data to be collected in real time and changes how we plan, govern, and operate cities (Chourabi et al. 2012). The expectations are that cities will be able to become "smarter in the long-term by continuous reflection in the short-term" (Batty 2012, p. 4), continuously anticipating the dynamics and changes. The big data generated by new technologies offer absolutely new innovative ways not only for service delivery, but also for better understanding of the cities themselves (Batty 2012). The ongoing collection of real-time geographical data on the city allows us to uncover new features of the interactions that take place in a city. As the services are improved throughout the cities, this leads to better quality of life and more efficient operations of the city. Ultimately, smart cities will be more competitive cities (Batty 2011).

Digital information substitutes and complements some elements of the physical infrastructure such as energy and materials (Batty 2012). Telecommuting might substitute for some of the actual commuting, while real-time traffic data analysis can provide for better transportation planning and reduction in commuting times. Another way digital data can complement the existing infrastructure is through better integration of existing services, such as the integration of different transportation modes. Even though the multimodal trip planning is still in infancy (Batty 2012), there are notable example such as TriMet Map Trip Planner[11] (launched in 2011, one of the first U.S. transit agencies implementing such modules).

These theoretical insights have never been empirically tested because of all those promises of smart cities discussed earlier. When faced with the question of what is a smart city, we need to focus on how smart cities use the technology and data. Perhaps one of the reasons why the smart cities concept holds so many promises is because of the anticipated ways the new technology and innovation will affect the quality of life and sustainability.

Quality of life and sustainability

Unlike the smart cities concept itself, quality of life has been measured even though it is difficult to estimate (Albouy 2009), because quality of life contains attributes such as human health and individuals' satisfaction from living in a place. Usually, quality of life is measured either through amenities (Albouy 2009) or as level of income or access to services and resources. Quality of life may also include certain attributes of cleanliness and aesthetic appeal of locations. Through innovation, smart cities improve quality of life and, thus, increase their competitive edge.

Since quality of life incorporates health and subjective characteristics of the cities, it is also impacted by the environmental conditions in the location. Sustainability is another difficult term to define even though there exists multiple various definitions, and the United Nations Environmental Programme (UNEP 2012, 2013) declares that green economy is sustainable economy. Allen and Clouth (2012, p. 8) use "non-declining human welfare over time" as the definition of sustainable development. That is, overfishing or pollution decreases the welfare of future generations. By this account, sustainable development involves paying close attention to externalities or when utility of individual A depends upon activities under his/her control and at least one activity under control of individual B (Buchanan and Stubblebine 1962), including externalities that involve intergenerational transfers (that is, future generations paying for the consumption of the current generation). The main focus becomes in internalizing environmental externalities and paying attention to equity issues (Allen and Clouth 2012). Reducing waste, pollution, and resource degradation become some of the important goals of sustainable development, and the part of sustainability emphasized for smart cities. UNEP (2012) coins the term "natural capital" similar to human capital or social capital to emphasize that natural resources can be viewed as a form of capital. The same way human capital captures more general characteristics of labor, natural capital is supposed to capture not only resources themselves but also surrounding services such as recycling, waste management, and so forth. However, having the goals of decreasing environmental externalities is not exactly the same as achieving those goals or understanding the mechanism underpinning the phenomena without proper measurements.

Unlike sustainability, the green economy has been measured in some studies and may not necessarily mean the same thing.

Green economy

As with other concepts discussed in this chapter, the green economy has become a cure for many ailments that include economic recovery, poverty eradication, reduced carbon emissions, and ecosystems' degradation (Allen and Clouth 2012; Green Economic Initiative UNEP). Discussions of "sustainable economic development" have increasingly focused on the development and promotion of so-called "green economies" (van der Bergh 2007). Even though the concept has long been recognized (e.g., Pearce 1989), UNEP defines a green economy as "one that results in improved human well-being and social equity, while significantly reducing environmental risks and ecological scarcities" (Sukhdev et al. 2010; UNEP 2011). In the United States, there has been an increasing interest in the economic and environmental benefits of transitioning to a "green economy" not only as a wealth-generation strategy but also as part of an effort to mitigate the effects of climate change (Muro et al. 2011). More prosaically, a recent study by the U.S. Department of Commerce defines the "green economy" as one in which goods are manufactured and services are provided in a way that conserves energy and other natural resources, furthers ecosystem conservation, and reduces pollution (U.S. Department of Commerce 2010). Given the importance of metropolitan areas in the economic life of the United States – they generate upward of 90 percent of all economic output and are home to over 84 percent of the nation's population – the development of a green economy in the United States will in effect require the country's urban economies to become "green" (Rees 1996).

Almost by definition, transitioning to a green economy implies the creation of green jobs (UNEP 2008). The challenge of the U.S. economy in becoming a green economy can therefore be posed quite specifically: U.S. urban economies, which together constitute the national economy, must create "green'" jobs with which to replace existing "nongreen" jobs.

Much of the focus of research on the green economy has been on green jobs (Muro et al. 2011; Pollack 2012; UNCSD 2011). What exactly is meant by "green jobs"? Although no standard definition for green jobs has been developed or accepted, many different definitions have been proposed by various organizations. The International Labour Organization (ILO) defines green jobs as those which improve energy and raw materials efficiency, limit greenhouse gas emissions, minimize waste and pollution, protect and restore ecosystems, and support adaptation to the effects of climate change (ILO 2013). In the United States, the Bureau of Labor Statistics and the Brookings Institution have engaged in significant efforts to define and count green jobs (Brookings Institution 2011; Department of Commerce 2010; EPI 2012). Typically, discussions and analysis of green jobs involve some kind of environmental component. Specifically, the focus is on occupations involving renewable energy generation and technology, energy efficiency, pollution prevention and cleanup, and natural resource conservation. Green jobs has been given a standard definition by the Occupational Information Network (O*NET), a program of the U.S. Department of Labor/Employment and Training Administration. O*NET focused on identifying green economic sectors, occupations with increased demand from green activities, green-enhanced skills–based occupations, and green new and emerging occupations (O*NET 2009). Of the 840 individual occupations identified by the U.S. Bureau of Labor Statistics (BLS), O*NET designated a subset of 144 occupations as green occupations. These occupations are those related to reducing the use of fossil fuels, decreasing pollution and greenhouse gas emissions, increasing the efficiency of energy usage, recycling materials, and developing and adopting renewable sources of energy (O*NET 2013).

The current debate is whether the green economy is the economy of the future which could have faster economic growth than other traditional sectors (Allen and Clouth 2012; Muro et al. 2011; Pollack 2012; UNCSD 2011). Although the green economy and green jobs are not easily identified and measured (Muro et al. 2011), recent BLS estimates for the green economy show that greener industries have grown faster, and that states with high green employment intensity faced less negative impact during the recent economic recession (Pollack 2012). UNEP (2012) argues that international trade can help to spread the environmentally friendly practices. Assuming cities are the basis for development across the board, sustainable development is a city-focused activity requiring new environmentally aware products and practices and new governance structures (UNCSD 2012).

What also unites all these concepts is the ultimate promise of economic development. Smart cities become laboratories for experimentation and innovation, and as such lead to economic development. Innovation can also be measured as patenting activity (Rothwell et al. 2013; Strumsky et al. 2011) or as creative class (Lobo et al. 2012). And large proportions of creative class in the regional economy can lead to economic development (Stolarick and Currid-Halkett 2012).

Creative class

A large number of different types of interactions happening in the same place increase the likelihood of innovations (Batty 2013; Bettencourt 2013). This leads us to the importance of social and human capital for economic development. Human capital has been shown to be highly correlated with increase in employment which may happen through increases in productivity and quality of life (Shapiro 2006).[12] The human capital is quite often measured by the number of people with university education (Albouy 2009; Glaeser et al. 2010; Glaeser and Resseger 2009; Glaeser and Saiz 2003; Shapiro 2006). Florida et al. (2008) indicate that creative class becomes a better measure of human capital as it allows us to study how various occupations can lead to improvements in technology which in turn improves productivity. Stolarick and Currid-Halkett (2012) find that cities with high creative class proportions did not experience the same unemployment rates as other areas during the most recent financial crisis.

The creative class influences productivity through technology (Florida et al. 2008). And technology is at the core of the smart cities concept (e.g., through real-time data analysis applications to city services) and green economy (e.g., through the innovations in the greener production processes). Unlike the previous concepts, the evidence has shown that creative class directly influences innovation (Stolarick et al. 2011), which improves city competitiveness (Batty 2011). Creative class also affects regional wages (Florida et al. 2012).

At the same time, green jobs are not necessarily "creative jobs" because a large portion of those jobs do not require college degrees (Pollack 2012), but these jobs still pay well to low- and middle-skilled workers (Muro et al. 2011). The lack of timely and objective data has complicated "the design of smart, realistic training and economic development systems at the regional level" (Muro et al. 2011, p. 42).

Smart, green, and creative concepts are ultimately intertwined conceptually. One of the first important features of smart cities is technology. The technology increases accessibility (through decreased transportation and communication costs), efficiency, and productivity. The technology is closely related to innovations. The innovations are also a component of creative class as Lobo et al. (2012) find the share of workforce engaged in creative activities as a good measure of regional inventive capacity. Because individuals create new ideas (Strumsky and Lobo 2012), creative class also influences the implementation and usage of technology in a smart city. Florida et al.'s (2008) research shows that consumer service amenities, among other factors such as tolerance, attract talent. Quality of life and technology are also affected by green economy. For example, the smart grid project (Smartgrid.gov) allows two-way communication between utilities and customers for more efficient energy use (Muro et al. 2011), and becomes an example of a component of both smart cities and green economy.

The uniting concept between smart cities, creative class, and green economy is the economies of agglomeration. Larger cities provide more benefits to creative occupations (Florida et al. 2012), and local amenities attract talent and creative class (Florida et al. 2008).

Agglomeration: Cities as networks

Bettencourt (2013) has proposed a new theory of the city that stems from the human interactions or understanding that all economic activity happens through exchanges. This application of "allometric laws" (Batty 2013) relates the increases in population size to scaling of the city attributes such as innovation activity (Lobo et al. 2013), crimes or wealth (Bettencourt et al. 2010). This new theory studies the internal dynamics of cities as networks. There are four major components of the theory (Bettencourt 2013). Minimal resources that can be accessed in a city are proportional to the costs of reaching those resources. The networks grow incrementally. Human effort is bounded. And socioeconomic output is proportional to the local social interactions. That is, because the cities are concentrations of social interactions, the declining transportation costs will not make cities less important. The agglomeration effects still work as cities become major exchanges for ideas and knowledge sharing (Batty 2013; Glaeser 2005; Glaeser and Pozetto 2010). Some attributes of cities scale superlinearly or exhibit economies of scale such as "production of patents, financial services, and crime" (Batty 2013, p. 1418); that is, innovative capacity of cities increases more than in proportion to population growth. Transportation, on the other hand, displays sublinear scaling with a city's population size or diseconomies of scale.

Spatial agglomeration states that the whole is larger than the sum of individual parts (Feser 1998). Marshall was the first who developed the concept of internal and external economies of scale and introduced spatial concentration of industries and firms (Feser 1998; Krugman 1991; Marshal 1890). The declining costs of communication and transportation do not undermine the economies of agglomeration because cities become places where exchange of ideas takes place (Batty 2012; Glaeser 2005). Agglomeration economies make interactions easier and bring new ideas to life (Batty 2012; Strumsky and Lobo 2012). Batty (2011) emphasizes that the old-fashioned view of cities as in the state of equilibrium does not apply; cities generate surprise. The cities are comprised of large networks of various kinds that make interactions between humans possible; thus, even when transportation costs decline, the large cities become innovative places due to agglomeration effects or the increased potential network connections (Batty 2012, 2013; Bettencourt 2013). The cities are representations of various networks (Batty 2007, 2011), and the challenge is to understand how to couple together material and "ethereal" (information) networks.

The cities become the ultimate innovation laboratories of new technology and policy experiments (Greenstein 2007; Hernandez-Munoz et al. 2011). Quality of life or livability is usually better in larger cities (Albouy 2008).

The agglomeration effects (or costs savings due to spatial proximity) can be generally related to the concept of externality, or the idea that actions of one influence the outcomes of another (Buchanan and Stubblebine 1962). The decisions of people to locate close to one another influence not only their own households, but also their neighbors, and, ultimately, increases the likelihood of various interactions. And because cities are typically viewed as networks, there is also the network effect, which represents the cost savings due to overall increase of the network or new additional agents added to the network. Bettencourt (2013) connects it more formally to the population size increase. Figure 6.1 depicts how various concepts of agglomeration relate to network externality. Overall, new theory connects network externality with spatial agglomeration. The spatial proximity of economic agents can be subdivided into localization and urbanization effects (Feser 2008). Localization economies are cost savings as the result of spatial concentration of production in the same industry. Localization is internal to industries, but external to firms (Feser 1998). Urbanization economies create cost savings because of the close proximity of various economic entities, increasing complementary knowledge and cross-product returns. In other words, localization is clustering of similar companies, while urbanization is clustering of many different services.

Marshall (1890) has separated the economies of scale into the external and internal economies. The external economies of scale occur when industry grows in size and generates costs savings for each individual firm, for example, through available labor pools in metropolitan areas. The internal economies of scale[13] represent cost savings to the firm size (Feser 1998). Glaeser (2010) differentiates the external economies into dynamic and static. The dynamic externalities increase the speed with which growth occurs or ideas are generated (Glaeser 2000) or represent superlinear scaling effects studied by Bettencourt (2013), Batty (2011), Bettencourt et al. (2010), and Lobo et al. (2013).

The external economies of scale can be due to spillover effects from advances in either knowledge or technology (Marshall 1890). Hoover (1948) has separated external economies in vertical and horizontal linkages. Horizontally linked companies tend to crowd out each other

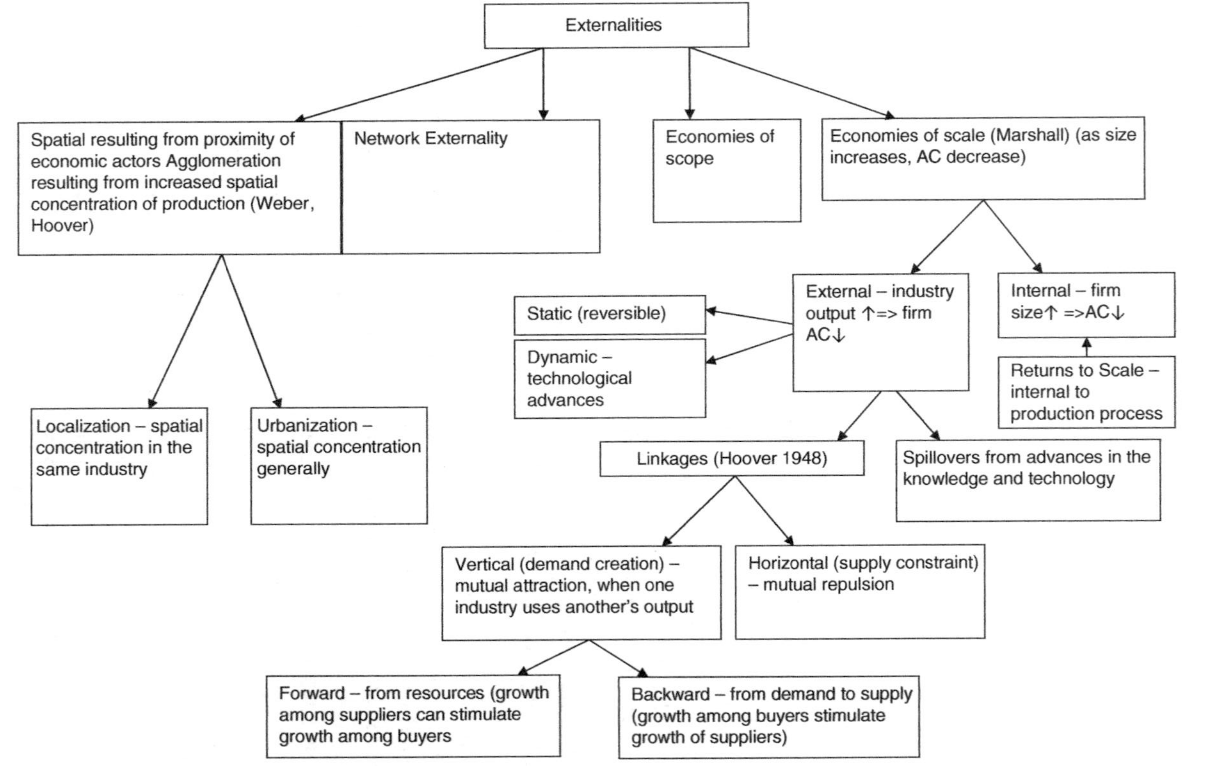

Figure 6.1 Externality, external economies, and related concepts.

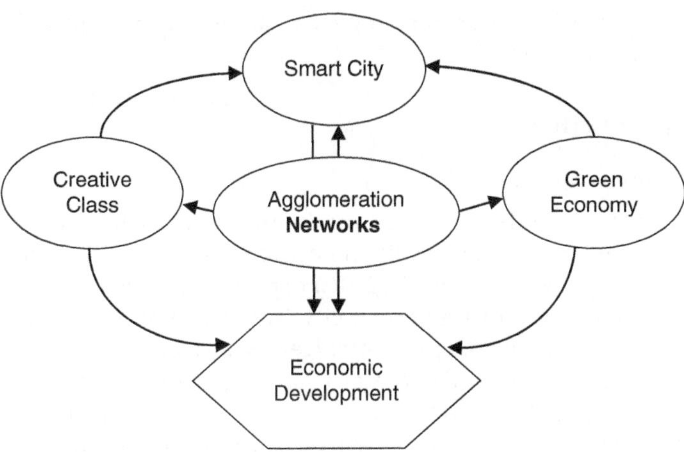

Figure 6.2 Agglomeration, smart cities, creative class, green economy, and economic development.

because of the high competition (Feser 1998). Vertically linked companies can benefit from agglomerating together through forward or backward linkages of supply chains. Logistics Performance Index indicates a gradual integration of different countries into global value chains.[14] That is, it is not simply that smart, green, and creative can be one and the same (even though they share common aspirations), but they all locate in the cities, and may provide their promises through individual human interactions.

Figure 6.2 connects agglomeration, economic development, and the concepts of smart cities, creative class, and green economy more explicitly.

At the core are agglomeration effects where spatial proximity of humans (or their interactions) (Bettencourt 2013) can lead to greater innovations (Glaeser and Gottlieb 2009; Lobo et al. 2013) and productivity (Batty 2013; Bettencourt 2013). Agglomeration ties all the themes together, as larger cities provide higher benefits for the creative class (Florida et al. 2012). Even as clean economy jobs concentrate in the largest metropolitan areas (Muro et al. 2011), the smart cities concept is supposed to decrease the costs of maintaining the city through innovative use of ICT. Moreover, all these concepts ultimately should lead to economic development, even though there is not much empirical evidence that they do in reality. One of the large problems is how to measure such big concepts. We discuss our innovative approach to smart cities and other measurements.

Measuring cities

Measuring smart cities

Despite a large number of studies on the subject of smart cities, there is a small number of operationalizable definitions of what the smart city concept entails. We use Batty's (2012) definition where ICT become an integral component of traditional city infrastructure. At the forefront where this can be measured is Intelligent Transportation System (ITS) deployment. The ITS incorporates a large spectrum of technologies, including automated vehicle location (AVL), computer-aided dispatch (CAD), environmental sensor stations (ESS), and dynamic message signs (DMS), to name a few. We use the 2010 ITS deployment tracking survey. The individual respondents are agencies that actually deploy ITS in various areas such as freeway management, arterial management, transportation management centers, transit management, fire and rescue, law enforcement, and toll collection. Only agencies in the largest metropolitan areas are surveyed. The response rate is anywhere from 81 percent to 93 percent depending on the area surveyed. The smart city score (*SC*) is calculated by summing up individual indexes by area (i), using the following formula:

$$SCi = \Sigma(AMi, FMi, TMi, TMCi, LEi, FRi, ETi) \tag{1}$$

For many areas, state-level agencies provide the most innovation in the ITS deployment. To account for this, we have calculated the state-level scores, and assigned them to metro and micro areas based on the locations' population.[15] This includes 33 state-level agencies recorded in the database. This assignment implies that locations receive the benefits of state-level ITS implementations.

Finally, we have calculated a location quotient for each metropolitan statistical area (MSA), according to the following formula:

$$LQmsai = \frac{aveSC/totalaveSC}{SCi/totalSCi} \tag{2}$$

That is, the numerator represents the proportion of MSA from the total smart score adjusted by the number of agencies reported, and the denominator represents the proportion of MSA smart score from the total MSAs smart number. The smart score LQs above 1 will indicate the situation where the share of MSA in average smart score is larger than the MSA's share in the total smart score. This indicates that when we adjust for the number of agencies operating and deploying ITS in the area, those areas with high LQ have more ITS deployments per agency. This ensures that the smart scores are not simply a function of a larger population as larger areas will have more agencies reporting.

It is important to note that even though this new approach to measuring smart cities is an innovation that can help us to empirically test the smart cities concept, it comes with certain limitations. For example, some places (e.g., San Francisco[16]) create useful web applications to browse the legal or travel data, and our measures do not capture this aspect. Our measure can also omit some of the newest pilot projects in use, for example, parking apps in some places[17] or collaboration between public and private entities on data sharing. At the same time, our measure captures the actual implementation of the technologies on the ground for traffic monitoring, safety improvements, and public transit planning.

Measuring sustainability

Project Vulcan[18] provides detailed information about CO_2 production at the metropolitan level across the United States. CO_2 emissions correlate with metropolitan gross metropolitan product (GMP) very positively and strongly at 0.808. However, it is not a perfect correlation. Figure 6.3 shows the scatterplot between logged emissions and logged GMP. The solid lines show the means of each and the dotted line is the best fit. So, as the correlation shows, while there is a strong linear relationship between the two, there are some interesting differences. Metros above the dotted line are producing greater CO_2 emissions for their GMP level than the average U.S. metro while those below the dotted line are producing fewer emissions for their GMP.

A sustainability index is constructed by doing the following:

1 Calculate the distance between the estimated line and the actual value. If a metro is below the line, use the distance. If a metro is above the line, multiply the distance by –1. Places with less-than-expected CO_2 emissions have a positive value while those with higher emissions are given a negative value.
2 Weight the value by the log of emissions. This weighting rewards cities for being both further from the line and also from having higher values.

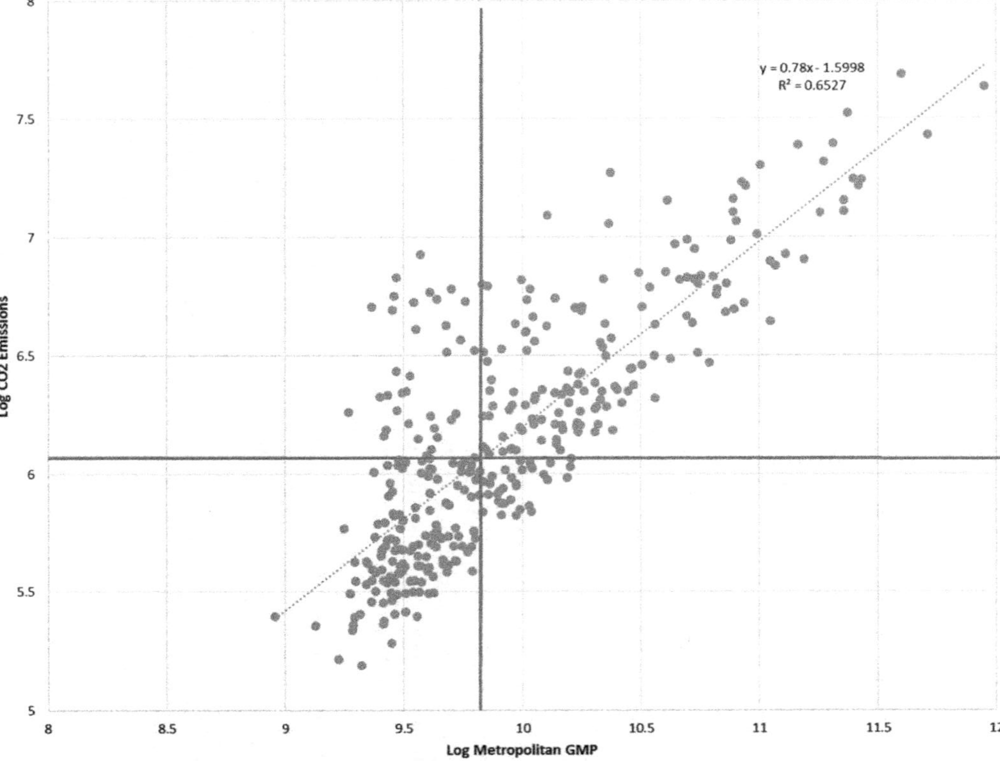

Figure 6.3 Gross metropolitan product and CO_2 emissions.

Table 6.1 Top ten/bottom ten sustainability index metros.

Metropolitan Area	CO_2 Emissions	GMP ($M)	CO_2 per $M GMP	Sustainability Index
Corvallis, OR (MSA)	190,648	2,798	68	2.033
Carson City, NV (MSA)	154,002	2,104	73	1.982
Bremerton-Silverdale, WA (MSA)	385,994	6,167	63	1.981
Jacksonville, NC (MSA)	248,744	3,587	69	1.944
San Jose–Sunnyvale–Santa Clara, CA (MSA)	4,425,378	111,512	40	1.944
Sioux Falls, SD (MSA)	687,329	10,953	63	1.812
Greenville, NC (MSA)	311,978	4,223	74	1.793
Bridgeport-Stamford-Norwalk, CT (MSA)	2,946,410	61,602	48	1.768
Bend, OR (MSA)	311,429	4,024	77	1.725
Manchester-Nashua, NH (MSA)	965,816	15,545	62	1.720
...	...	...	...	...
Parkersburg-Marietta-Vienna, WV-OH (MSA)	5,491,287	4,361	1,259	–4.362
Baton Rouge, LA (MSA)	18,546,305	23,513	789	–4.462
Evansville, IN-KY (MSA)	12,351,663	12,777	967	–4.522
Terre Haute, IN (MSA)	5,857,583	4,091	1,432	–4.645
Monroe, MI (MSA)	5,328,128	3,497	1,524	–4.681
Manhattan, KS (MSA)	4,928,409	2,832	1,740	–4.855
Flagstaff, AZ (MSA)	5,631,886	2,864	1,966	–5.186
Pine Bluff, AR (MSA)	5,068,480	2,312	2,192	–5.292
Weirton-Steubenville, WV-OH (MSA)	6,723,406	2,931	2,294	–5.617
Farmington, NM (MSA)	8,453,731	3,720	2,273	–5.801

Using this index, the top ten/bottom ten metros (among the 361) are shown in Table 6.1. Not unexpectedly, those metros with the highest sustainability index produce less than one-twentieth the CO_2 per dollar of GMP than those with the lowest sustainability index (average 63.7 versus 1,643.5).

Other Measures

The other measures used in this analysis are from existing sources and are well documented elsewhere. These measures include the following:

- Green Jobs – from the definition used by the Brookings Institution in its report on the green economy (Muro et al. 2011) and based on BLS occupational data – location quotient (concentration) of jobs in the "green" category compared to the national average.
- Creative Class and Super Creative Core – based on the Florida (2008) definition and using data from the BLS Occupational Employment Statistics – also a location quotient. Results for the Creative Class and the Super Creative Core are very similar, and only Creative Class relationships will be discussed.
- GMP – regional gross domestic product (GDP) estimate from the Bureau of Economic Analysis.
- Regional Employment – size of regional workforce – total number from Occupational Employment Statistics (OES) BLS.

Results

Data Overview

Table 6.2 shows the summary statistics for the various measures used in this analysis. The number of metropolitan areas for which data is available varies by measure. For green jobs variable both metropolitan and micropolitan area data is available, but only data from metropolitan areas has been used in this analysis. For the smart cities measure, some scores were assigned from the state-level smart scores. In addition to the standard statistics, the values for the 40 percent (second quintile) and 60 percent (fourth quintile) are also shown. These values will be used later to split the metros for each measure to determine whether a city is in high (4th and 5th quintile) or low (1st or 2nd quintile) ranges for multiple measures.

Table 6.3 shows the pairwise correlations for the various measures. The correlations show both expected, unexpected, and exploratory relationships. The expected relationships are the strong positive correlations among GMP (productivity), total employment (agglomeration), and Creative Class. All of these measures have fairly strong correlations. The productivity/agglomeration relationship is a well-documented and expected one (Batty 2013; Bettencourt 2013). The Creative Class and productivity and agglomeration relationships have also been previously reported (Florida et al. 2012).

The more unexpected relationship is the negative and fairly strong correlation between the Smart Cities LQ and productivity and agglomeration.

That greater investment in smart transportation infrastructure is more likely in smaller metros is not necessarily a surprise – smaller regions would find it less expensive to implement these technologies, and benefit from the state-level efforts although the possibility of economies of scale would counter that argument. The greater surprise is the very

Table 6.2 Summary statistics.

	N	Min	Max	Average	Std Dev	Median	Quintile 1 & 2 (40%)	Quintile 4 & 5 (60%)
Green Jobs LQ	366	0.160	5.378	1.075	0.756	0.906	0.813	1.014
Sustainable Index	361	0.002	0.837	0.181	0.147	0.150	0.123	0.180
Smart Cities LQ	287	0.127	7.910	5.115	3.006	6.989	3.955	7.910
Creative Class LQ	345	0.629	1.617	1.005	0.160	0.986	0.950	1.018
Super Creative Core LQ	345	0.166	2.935	1.005	0.290	0.959	0.905	1.006
Log GMP per capita	361	4.243	7.007	5.229	0.473	5.133	4.989	5.244
Log Total Employment	361	4.520	5.032	4.715	0.078	4.710	4.690	4.726

Table 6.3 Correlations.

	Sustainability Index	Smart Cities LQ	Creative Class LQ	Super Creative Core LQ	Log GMP per capita	Log Total Employment
Green Jobs LQ	–0.01	0.21	0.10	0.07	–0.04	0.09
Sustainability Index		0.09	–0.15	–0.10	–0.21	0.01
Smart Cities LQ			–0.28	–0.13	–0.65	–0.40
Creative Class LQ				0.84	0.49	0.42
Super Creative Core LQ					0.30	0.23
Log GMP per capita						0.68

strong and negative correlation between GMP per capita (productivity) and smart cities. It is possible that is an anti-selection bias – places with lower productivity are more likely to invest in smart cities in an attempt to improve productivity. Whether that is the case, this is definitely an interesting finding and something that should be investigated in more detail in future research.

The more exploratory relationships are those among green jobs, the Sustainability Index, the Smart Cities LQ, and the Creative Class. Essentially, the results show that no significant positive or negative correlations were found for these measures. This includes the finding that there is no relationship between green jobs and the Sustainability Index. Essentially, places that have a higher concentration of people working in green jobs are no more or less likely to be producing lower CO_2 emissions based on the regional GMP. Neither green jobs nor sustainability is more or less likely when a region has implemented smart transportation initiatives. And regional productivity and agglomeration do not seem to be associated with either green jobs or sustainability. The Creative Class concentration in region is also not higher in regions with a higher concentration of green jobs, greater sustainability, or smart cities initiatives. Overall, the correlation results suggest that metros that have green jobs or are sustainable (and those are not the same thing) are neither smart nor creative. Smart is not creative. And green jobs, sustainability, or degree of smart transportation initiatives are not higher in regions with higher productivity or agglomeration.

In the next section we will look at an alternative way of considering this relationship. We will then conclude with a brief discussion.

Smart, Green, and Creative?

We employ the Venn diagrams to capture all possible intersections of our three concepts. Figure 6.4 represents a hypothetical Venn diagram, showing smart cities with high proportion of Creative Class, smart cities with high proportion of green jobs, creative places with large proportions of green jobs, and finally places with all three characteristics being high. Because all our variables are measured as Location Quotients (or similar metric for smart cities), the highest values will correspond to the highest rank.

Muro et al. (2011) indicate that there might not be too much intersection between Creative Class and green jobs, so we might find situations like in Figure 6.5 where Creative Class and green jobs do not have much correlations with each other. Or they may not overlap at all.

The correlation results suggest that on an overall basis, little association exists among the three concepts using various measures. However,

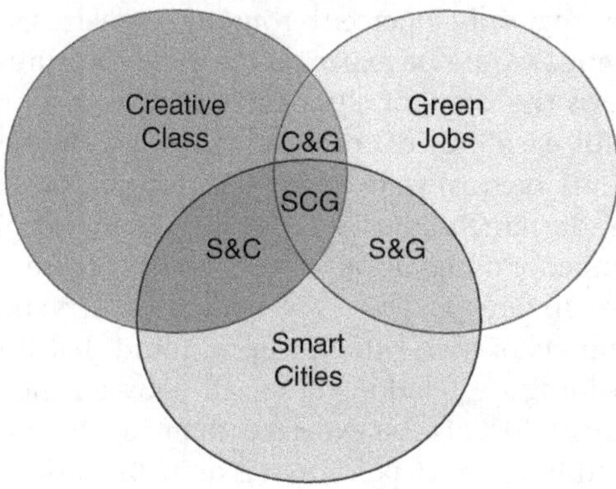

Figure 6.4 A hypothetical Venn diagram with high proportions of smart, green, and creative cities.

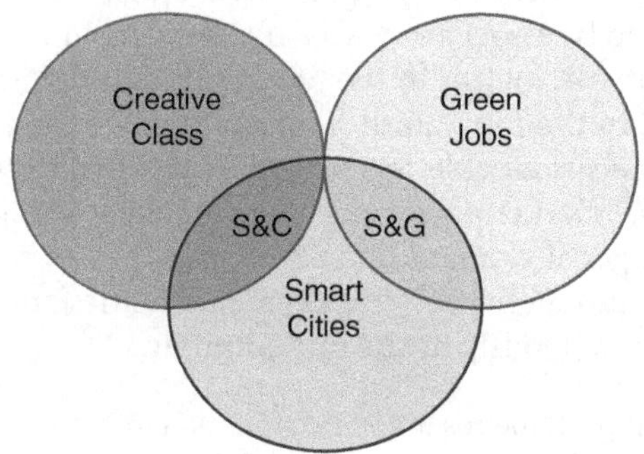

gure 6.5 A hypothetical Venn diagram where creative class cities and green ›bs places do not have overlap.

that does not mean that a relationship may not be possible but only exists within more narrowly defined categories. For example, if only a small number of cities are simultaneously green, smart, and creative while others are green and smart but not creative and others are smart and creative but not green, and so forth, the correlations may not be positive and significant. To test for this, we divided each measurement into quintiles and then identified if a metro was in the high (top two: quintile 4 and 5) or low (bottom two: quintile 1 and 2) for each measure. We could then calculate the share of all metros that were in the high or low quintiles for multiple measures. (The results shown are for quintiles 4 and 5, and 1 and 2. Results using only the highest, quintile 5, and lowest, quintile 1, were also done and were consistent with the results presented here but were less robust because using only one quintile for two measures meant that only 4 percent, roughly 15 metros, and less than 1 percent or 3 metros would be expected for three measures.)

Table 6.4 shows the share of all possible metros that are in the high/high category for each pair of measures. So, for example, 18.3 percent (actually 66 of 361 metros) were in the top two quintiles for both green jobs LQ and the Sustainability Index. The shares are calculated based on the smaller number of observations for each pair because only that many paired observations are possible. The various observations across the measures are subsets of each other. Keep in mind that if independently distributed, meaning no relationship at all between the two measures, 16 percent (40% of 40%) is the expected share. So, when the calculated share is significantly above 16 percent, it is more likely than chance would suggest that a higher score in one measure is associated with a higher score in another measure for a subset of metro regions. If the calculated share is significantly below 16 percent, then a higher score for one measure is less likely to be associated with a higher score in the other measure.

Looking at Table 4, metros in the top quintiles for green jobs are more likely to also be in the top quintiles for smart cities and are weakly associated with more sustainable regions, more creative regions, and larger agglomerations. Smart cities are less likely to be in the higher productivity quintiles (larger agglomerations). And, as expected from the correlations, Creative, productive, and agglomerated cities are all much more likely to be simultaneously in the top quintiles.

Table 6.4 Percent of all metros in top two quintiles in both measures.

HIGH/ HIGH	Sustainability Index	Smart Cities LQ	Creative Class LQ	Super Creative Core LQ	Log GMP per capita	Log Total Employment
Green Jobs LQ	18.3%	23.7%	17.7%	17.1%	15.2%	18.3%
Sustainability Index		16.3%	13.3%	14.5%	12.7%	15.8%
Smart Cities LQ			16.0%	18.8%	10.1%	14.3%
Creative Class LQ				31.6%	26.4%	23.5%
Super Creative Core LQ					22.3%	20.6%
Log GMP per capita						27.4%

Table 6.5 shows the other end of the scale – metros that are in the lowest quintiles for each of the pairs. So, for example, 15.2 percent of metros (55 of 361) are in lowest quintiles for green jobs and sustainability. Again, 16 percent is the expected value and the interest is in values significantly above or below 16 percent. Generally, the results in Table 6.5 mirror those in Table 6.4 – metros that were likely to have high/high pairs also are more likely to have low/low pairs. This again is especially strong for Creative, productivity, and agglomeration. Metros with a low smart cities score are also less likely to have a low productivity or agglomeration score. The Sustainability Index is not that different from random chance showing little relationship with the other variables. Metros with a low green jobs LQ are slightly more likely to also have a low smart cities LQ and slightly more likely to have a lower Creativity, productivity, or agglomeration score.

The final consideration is to look at the tuple of measures. In this case, the expected score would be 6.4 percent (40% of 40% of 40%), and the number of metros for which each tuple of scores is available is 287. The first thing is to consider if metros with green jobs are more likely to also be smart and creative. The results are as follow:

Green – Smart – Creative
HHH = 23/287 = 8.0%
LLL = 22/287 = 7.7%

Table 6.5 Percent of all metros in bottom two quintiles in both measures.

LOW/ LOW	Sustainability Index	Smart Cities LQ	Creative Class LQ	Super Creative Core LQ	Log GMP per capita	Log Total Employment
Green Jobs LQ	15.2%	19.9%	17.4%	17.4%	18.3%	20.8%
Sustainability Index		15.5%	14.5%	15.4%	13.3%	16.6%
Smart Cities LQ			12.2%	13.6%	9.1%	13.2%
Creative Class LQ				30.7%	24.9%	22.6%
Super Creative Core LQ					22.9%	19.7%
Log GMP per capita						27.4%

So, for 23 of the 287 metros (8.0%), which is more than the 18 expected at random but not very much more, are ones in the top two quintiles for green jobs AND smart cities AND Creative Class. The number of metros for which the three measures are all in the lowest quintiles is 22, which is just 4 more than the expected number. If these three measures were perfectly correlated, 115 metros would be in the joint highest category and an equal number would be in the joint lowest. Given these results, it would be a major stretch to suggest that green metros are also smart and creative.

Table 6.6 shows the 23 metros in the highest quintiles and the 22 metros in the lowest quintiles. Most of these metros have a Smart City score that is the result of being in a state that has implemented several intelligent transportation systems. While both lists of metros seem to make intuitive sense (Pueblo, CO, excepted), this is clearly a case of simply picking winners and losers. So, while this is a list of metros that *are* green, smart, and creative, the analysis shows that this is not a statistically significant relationship.

Table 6.6 Green-smart-creative cities.

High-High-High	Low-Low-Low
Ames, IA	Alexandria, LA
Anchorage, AK	Bakersfield-Delano, CA
Athens-Clarke County, GA	Brownsville-Harlingen, TX
Augusta-Richmond County, GA-SC	Cape Coral-Fort Myers, FL
Bismarck, ND	Corpus Christi, TX
Bloomington, IN	Deltona-Daytona Beach-Ormond Beach, FL
Cheyenne, WY	Houma-Bayou Cane-Thibodaux, LA
Chico, CA	Kankakee-Bradley, IL
Columbus, GA-AL	Lafayette, LA
Durham-Chapel Hill, NC	Laredo, TX
Goldsboro, NC	Las Vegas-Paradise, NV
Huntsville, AL	McAllen-Edinburg-Mission, TX
Idaho Falls, ID	Midland, TX
Kennewick-Pasco-Richland, WA	Odessa, TX
Knoxville, TN	Owensboro, KY
Kokomo, IN	Pueblo, CO
La Crosse, WI-MN	Sandusky, OH
Lafayette, IN	Shreveport-Bossier City, LA
Madison, WI	Springfield, OH
Missoula, MT	Texarkana, TX-Texarkana, AR
Olympia, WA	Victoria, TX
Redding, CA	Wichita Falls, TX
Santa Cruz-Watsonville, CA	

However, an alternative measure of regional "greenness," the Sustainability Index, was also created. The results below consider regions that are simultaneously in the top quintiles or bottom quintiles for the sustainable, smart, creative tuples.

Sustainable – Smart – Creative

HHH = 21/287 = 7.3%

LLL = 13/287 = 4.5%

These results are even less encouraging. The share of metros that have scores in the highest quintiles is not much greater than what would be expected from an independent relationship. And, while the lower quintile scores is lower than random chance, it too is far from the 115 metros that would share the low scores if they were perfectly correlated.

Table 6.7 shows the 21 metros in the highest quintiles and the 13 metros in the lowest quintiles. As in Table 6.6, most of the metros have a Smart City score that is the result of being in a state that has implemented several intelligent transportation systems and not a Smart City score that is the direct result of actions taken within the metro. The ten lightly shaded metros are those that are also in the top quintiles for

Table 6.7 Sustainable-smart-creative cities.

High-High-High	**Low-Low-Low**
Bremerton-Silverdale, WA	Bellingham, WA
Columbus, GA-AL	Cape Coral-Fort Myers, FL
Columbus, IN	Deltona-Daytona Beach-Ormond Beach, FL
Durham-Chapel Hill, NC	Grand Junction, CO
Fayetteville, NC	Houma-Bayou Cane-Thibodaux, LA
Huntsville, AL	Kankakee-Bradley, IL
Idaho Falls, ID	Las Vegas-Paradise, NV
Iowa City, IA	McAllen-Edinburg-Mission, TX
Kennewick-Pasco-Richland, WA	Riverside-San Bernardino-Ontario, CA
Kokomo, IN	Scranton–Wilkes-Barre, PA
La Crosse, WI-MN	Springfield, OH
Macon, GA	Texarkana, TX-Texarkana, AR
Missoula, MT	Waco, TX
Morgantown, WV	
Olympia, WA	
Santa Cruz-Watsonville, CA	
Santa Rosa-Petaluma, CA	
State College, PA	
Tulsa, OK	
Warner Robins, GA	
Winston-Salem, NC	

Green. Those ten are places that are highly ranked among all metros for Green-Sustainable-Smart-AND-Creative. Table 6.7 also highlights the eight metros that score at the bottom of all four metrics. Again, both the high scoring and low scoring lists, while interesting and revealing about the specific metros listed, don't reveal anything about the overall relationships for U.S. metros.

Discussion and Conclusions

So, are green cities smart? Are they also creative? The results are pretty clear, and the answer is "no." From both the correlations and quintile analysis, it is clear that larger cities are more productive and are also places that have a larger concentration of creative workers. It can be said that there are some indications that places with a higher concentration of green jobs are also more likely to be places that have implemented more intelligent transportation systems. So, from that perspective, green cities can also be smart cities. But green cities are not creative cities. And if a green city is not measured by employment but by environmental impact from CO_2 emissions, then green/sustainable cities are neither smart nor creative. Smart cities are not creative cities. And green, sustainable, and smart cities are not more productive or larger agglomerations.

The results also show that few cities are simultaneously green, smart, and creative or sustainable, smart, and creative.

To return to our Venn diagrams, Figures 6.6 and 6.7 summarize our findings. Green cities can also be smart cities, but no other significant relationships were found among the other measures. The following Venn diagrams represent found relationships. Figure 6.7 – essentially

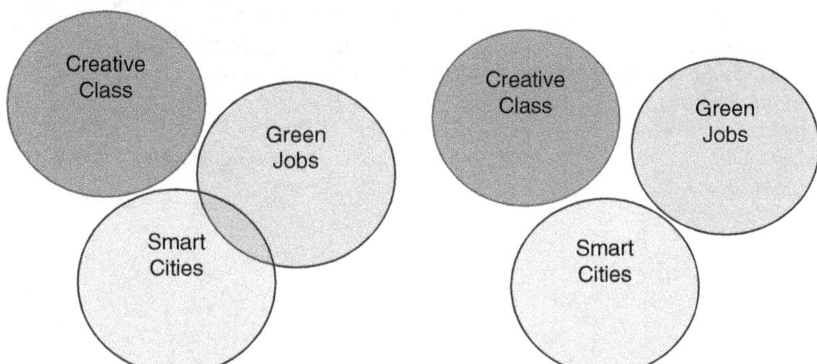

Figure 6.6 A Venn diagram based on findings: weak overlap between smart cities and green jobs locations.

Figure 6.7 A Venn diagram based on the current findings: no significant overlap between the locations.

showing no significant overlap among U.S. metropolitan regions – is the most accurate.

Much has been said about green cities, sustainable cities, smart cities, and creative cities. And much more is likely to be said. Using existing, previously developed measures for creative and green cities and developing new measures for sustainable and smart cities, we have shown that they are currently not the same. We started this exercise expecting that while we might not find incredibly strong relationships among the three kinds of metro regions – green/sustainable, smart, and creative – we did expect that there would be some commonalities. They simply are not there. These are different concepts, differently measured, and while each may have its own importance and value to many regions for various reasons, they are different dimensions that are quite independent of each other – at least currently and using the measures available.

While these are the best measures available, they clearly may not be capturing the nuance and diversity of programs and other efforts that define a region as being green or smart or creative. However, they do show that these are independent concepts that each must be developed and pursued on its own. A region will not be green or sustainable or smart simply by employing lots of Creative Class. Nor will being green or sustainable result in more Creative Class in the region. The strength of these results is showing that these are independent concepts that require independent policies.

Our findings might be a reflection that these terms are used by cities in the same way as firms use greening initiatives as "competitive strategic weapons" (Min and Kim 2012). Maybe our findings stem from the reality that both smart cities and green economy (or green jobs) are more goals or in their infancy of development than fully developed and realized urban development strategies, and we need more time before we will be able to see different results. Time will tell whether cities will indeed be able to become smarter, greener, and more sustainable by using the big data generated by new technologies or whether the cities will be simply overwhelmed by the new data.[19] The green economy may continue to grow at a faster rate than the regular economy or the current estimates of its growth might overestimate its effects.[20]

References

Albouy, D. (2008). "Are big cities bad places to live? Estimating quality of life across metropolitan areas." NBER working paper No. 14472, available online at http://www.nber.org/papers/w14472

Albouy, D. (2009). "What are cities worth? Land rents, local productivity, and the capitalization of amenity values." NBER working paper No. 14981, available online at http://www.nber.org/papers/w14981

Allen, C., and Clouth, S. (2012). "A guidebook to the green economy." Publication of the Division for Sustainable Development UNDESA, available online at http://sustainabledevelopment.un.org/content/documents/GE%20Guidebook.pdf

Batty, M. (2011). "Building a science of cities." Working Papers Series paper 170-Nov11, pp. 1–14, available online at https://www.bartlett.ucl.ac.uk/casa/pdf/paper170.pdf

Batty, M. (2012). "Smart cities of the future." Working Papers Series paper 188-Oct12, pp. 1–41, available online at https://www.bartlett.ucl.ac.uk/casa/pdf/paper188

Batty, M. (2013). "A theory of city size." *Science,* 340(6139), pp. 1418–1419.

Bell, C. (1988). "Economies of, versus Returns to, Scale: A Clarification," The Journal of Economic Education, vol. 19, issue 4, pp. 331–335.

Bettencourt, L. (2013). "The origins of scaling in cities." *Science,* 340(6139), pp. 1438–1441.

Bettencourt, L., Lobo, J., Strumsky, D., and West, G. (2010). "Urban scaling and its deviations: Revealing the structure of wealth, innovation and crime across cities." *PLoS ONE,* 5(11): e13541, available online at http://www.plosone.org/article/fetchObject.action?uri=info%3Adoi%2F10.1371%2Fjournal.pone.0013541&representation=PDF

Buchanan, J., and Stubblebine, C. (1962). "Externality." *Economica,* 29(116), pp. 371–384.

Chourabi, H., Gil-Garcia, R., Pardo, T., Nam, T., Mellouli, S., Scholl, H. J., Walker, S., and Nahon, K. (2012). "Understanding smart cities: An integrativeframework." *Proceedings of 2012 45th Hawaii International Conference on System Sciences,* available online at http://www.ctg.albany.edu/publications/journals/hicss_2012_smartcities/hicss_2012_smartcities.pdf

Feser, E. (1998). "Enterprises, external economies, and economic development." *Journal of Planning Literature,* 12(3), pp. 283–302.

Florida, R., Mellander, C., and Stolarick, K. (2008). "Inside the black box of regional development – human capital, the creative class, and tolerance." *Journal of Economic Geography,* 8(5), pp. 615–649.

Florida, R., Mellander, C., Stolarick, K., and Ross, A. (2012). "Cities, skills and wages." *Journal of Economic Geography,* 12, pp. 355–377.

Glaeser, E. 2000. "The new economics of urban and regional growth." In G. Clark, M. Feldman, and M Gertler, eds., *The Oxford Handbook of Economic Geography,* Oxford: Oxford University Press, pp. 83–98.

Glaeser, E. 2005. "Urban colossus: Why is New York America's largest city?" NBER working paper No. 11398, available online at http://www.nber.org/papers/w11398

Glaeser, E., and Gottlieb, J. (2009). "The wealth of cities: Agglomeration economies and spatial equilibrium in the United States." NBER working paper No. 14806, available online at http://www.nber.org/papers/w14806

Glaeser, E., and Ponzetto, G. (2010). "Did the death of distance hurt Detroit and help New York?" in Edward Glaeser, ed., *Agglomeration Economics,* Chicago: University of Chicago Press, pp. 303–337.

Glaeser, E., Ponzetto, G., and Tobio, K. (2011). "Cities, skills, and regional change." NBER working paper No. 16934, available online at http://www.nber.org/papers/w16934

Glaeser, E., and Resseger, M. (2009). "The complementarity between cities and skills" NBER working paper No. 15103, available online at http://www.nber.org/papers/w15103

Glaeser, E., and Saiz, A. (2003). "The rise of the skilled city." NBER working paper No. 10191, available online at http://www.nber.org/papers/w10191

Greenstein, S. (2007). "Economic experiments and neutrality in Internet access." In Adam B. Jaffe, Josh Lerner, and Scott Stern, eds., *Innovation Policy and the Economy*, 8, Cambridge, MA: NBER, pp. 59–109.

Gyourko, J., Mayer, C., and Sinai, T. (2010). "Dispersion in house price and income growth across markets." In Edward Glaeser, ed., *Agglomeration Economics*, Chicago: University of Chicago Press, pp. 67–104.

Hernandez-Munoz, J. M.,Vernat, J. B., Munoz, L., Galache, J. A., Presser, M., Hernandez Gomez, L. A., and Pettersson, J. (2011). "Smart cities at the forefront of the future Internet." In John Dominique et al., eds., *The Future Internet. Future Internet Assembly 2011: Achievements and Technological Promises*, pp. 447–462.

Hoover, E. M. (1948). *The location of economic activity.* New York: McGraw-Hill.

Krugman, P. (1991). *Geography and trade*. Cambridge, MA: MIT Press.

Lobo, J., Bettencourt, L., Strumsky, D., and West, G. (2013). "Urban scaling and the production function for cities." *PLoS ONE*, 8(3), p. e58407, available online at http://www.plosone.org/article/info%3Adoi%2F10.1371%2Fjournal.pone.0058407

Lobo, J., Mellander, C., Stolarick, K., and Strumsky, D. (2012). "The inventive, the educated, and the creative: How do they affect metropolitan productivity?" working paper #263 CESIS Electronic Working Paper Series, pp. 1–40, available online at http://www.kth.se/dokument/itm/cesis/CESISWP263.pdf

Lobo, J., Strumsky, D., and Rothwell, J. (2013). "Scaling of patenting with urban population size: Evidence from global metropolitan areas." *Scientometrics*, 96, pp. 819–828.

Marshall, A. 1890. *Principles of economics*, 8th ed. London: Macmillan.

Min, H., and Kim, I. (2012). "Green supply chain research: Past, present, and future." *Logistics Research*, 4(1–2), pp. 39–47.

Muro, M., Rothwell, J., and Saha, D., with Battelle Technology Partnership Practice. (2011). Sizing the Clean Economy Series, Brookings Institution. Available online at http://www.brookings.edu/research/reports/2011/07/13-clean-economy

Pollack, E. (2012). "Assessing the green economy and its implications for growth and equity." EPI Briefing Paper #349, available online at http://www.epi.org/publication/bp349-assessing-the-green-economy/

Rothwell, J., Lobo, J., and Strumsky, D. (2013). "The role of invention in U.S. metropolitan productivity." Social Science Research Network working paper series, available online at http://papers.ssrn.com/sol3/papers.cfm?abstract_id=2175310

Schaffers, H., Komninos, N., Pallot, M., Trousse, B., Nilsson, M., and Oliveira, A. (2011). "Smart cities and the future Internet: Towards cooperation frameworks for open innovation." In John Dominique et al., eds., *The Future Internet. Future Internet Assembly 2011: Achievements and Technological Promises*, pp. 431–446.

Shapiro, J. (2006). "Smart cities: Quality of life, productivity, and the growth effects of human capital." *The Review of Economics and Statistics*, 88(2), pp. 324–335.

Stolarick, K., and Currid-Halkett, E. (2012). "Creativity and the Ccisis: The impact of creative workers on regional unemployment." *Cities*, 33, pp. 5–14.

Stolarick, K., Lobo, J., and Strumsky, D. (2011). "Are creative metropolitan areas slso entrepreneurial?" *Regional Science Policy and Practice*, 3(3), pp. 271–287.

Strumsky, D., and Lobo, J. (2012). "Scale effects in invention? Evidence from urban areas in the U.S." Social Science Research Network working paper series, available online at http://papers.ssrn.com/sol3/papers.cfm?abstract_id=2194042

Strumsky, D., Lobo, J., and van der Leeuw, S. (2011). "Using patent technology codes to study technological change," Santa Fe Institute Working Paper 10-11-023.

UNCSD Secretariat (2011). "Green jobs and social inclusion." RIO 2012 Issues Brief #7, available online at https://sustainabledevelopment.un.org/sdissues briefs.html.

UNCSD Secretariat (2012). "Green economy knowledge – sharing platform: Exploring options." Information Note, available online at https://sustainable development.un.org/sdissuesbriefs.html.

UNEP (2012). "Valuing nature." Brief Paper on Green Economy, available online at http://www.unep.org/greeneconomy/Portals/88/VALUING%20NATURE.pdf

UNEP (2013). "Green economy and trade – trends, challenges and opportunities." Available online at http://www.unep.org/greeneconomy/Green EconomyandTrade

Notes

1 Special thanks to José Lobo for contributing his thinking during the development of this chapter. Also, we thank James Pol (DOT ITS Joint Program Office) and Stephen Gordon (ORNL) for their help with the access to the 2010 ITS deployment survey.
2 www.frost.com/prod/servlet/cpo/213016007
3 Chourabi et al. (2012) list multiple definitions of smart cities, but indicate that all of them include the importance and influence of technology.
4 For example, energy efficiency of the computers doubles every 18 months (http://www.technologyreview.com/news/425398/a-new-and-improved-moores-law/).
5 For more information, see http://www.ibm.com/smarterplanet/us/en/smarter_cities/overview/.
6 For additional information, see http://cities.media.mit.edu/.
7 For more information, see http://ec.europa.eu/eip/smartcities/index_en.htm.
8 "We shape our tools and therefore tools shape us" (Batty 2012 p. 5).
9 For more information, see http://smartcitiescouncil.com/.
10 http://www.fhwa.dot.gov/livability/
11 More information on TriMet, see http://trimet.org/howtoride/maptripplanner.htm, or on multimodal trip-planning, see http://www.vtpi.org/multimodal_planning.pdf.
12 Shapiro (2006) indicates that about 60% of employment growth happens due to productivity improvements, while the rest is likely due to quality-of-life improvements. Preliminary findings also indicate that quality-of-life improvements occur mainly through bars and restaurants.
13 There are subtle differences between the economies of scale and the returns to scale. The economies of scale capture decreases in cost per unit as firm size increases. Returns to scale are internal to the production process; that is, as

the inputs double, the output more than doubles for increasing returns to scale. Bell (1988) postulates the increasing returns to scale imply the economies of scale, but not vice versa. At the same time, a production might have decreasing returns to scale, but still exhibit average costs reductions per unit as the company size grows (or economies of scale).

14 http://people.hofstra.edu/geotrans/eng/ch5en/conc5en/LPI.html

15 For MSAs crossing state boundaries, we use the proportion of population in a state to assign state scores.

16 Specifically on San Francisco project, see http://sanfranciscocode.org/; for more on the law decoding, see http://americadecoded.org/.

17 For more information, see http://smartcitiescouncil.com/article/list-you-dont-want-be-cities-dirtiest-air.

18 For more information, see http://vulcan.project.asu.edu/.

19 There is a debate over whether current ITC developments have contributed to human productivity or have added more distractions from work.

20 For some alternative estimations of green economy growth at the global level, see http://www.nature.com/nature/journal/v472/n7343/full/472295a.html.

7

Urban Reconfiguration after the Emergence of Peer-to-Peer Infrastructure: Four Future Scenarios with an Impact on Smart Cities

Vasilis Kostakis, Michel Bauwens, and Vasilis Niaros

Introduction

Today, the majority of human beings are city dwellers. In this increasingly urbanized world, smart cities are emerging as an alternative city model to tackle several environmental, economical, and societal issues. Although there is not any compact and agreed-upon definition of smart cities, cities are generally defined as "smart" when they are infused with information and communication technologies (ICT), and a social infrastructure that promotes sustainability and active engagement of citizens (Caragliu, Del Bo, & Nijkamp, 2009). In the current environment, rapidly progressing ICT and the subsequent emergence of peer-to-peer (P2P) infrastructure are giving rise to potentially limitless innovation that can be implemented in cities to improve efficiency and connectivity.

To be more precise, P2P infrastructure is that infrastructure for communication, cooperation, and common value creation that allows for permission-less interlinking of human cooperators and their technological aids. We argue that such infrastructure is becoming the general condition of work, life, and society with the potential to reshape the idea of the "smart city." P2P relational dynamics, which epitomize the old slogan "Jeder nach seinen Fähigkeiten, jedem nach seinen Bedürfnissen!" (From each according to his ability, to each according to his need!), are based on the distribution of our productive forces.

First, the means of information, immaterial production (i.e., the networked computers) and now the means of physical, material production (i.e., machines that produce physical objects) are being distributed and interconnected. Just as networked computers democratized the means of production of information and communication, the emergent elements of networked microfactories or what some (see Kostakis, Fountouklis, & Drechsler, 2013) call desktop manufacturing, such as three-dimensional (3-D) printing, are democratizing the means of production.

Of course, this is not by any means an unproblematic process. In a period of extreme socioeconomic polarization and lacking any equilibrium regarding the global governance of the Internet (Mueller, 2010), we have been witnessing conflicts for the control and ownership of distributed infrastructure. On the one hand, commons-based peer production signals fundamental changes in value creation, especially when juxtaposed against an old order that is in decline (see Bauwens, 2005; Benkler, 2005; Kostakis, 2013). On the other hand, the proposed legislations of Anti Counterfeiting Trade Agreement (ACTA)/Stop Online Piracy Act (SOPA)/Protect IP Act (PIPA) enforce strict copyright within a regulatory regime that polices transactions beforehand instead of afterward (Boyle, 1997). Furthermore, the attempt for surveillance and censorship by both authoritarian and liberal countries, and "the growing tendency to link the Internet's security problems to the very properties that made it innovative and revolutionary in the first place" (Mueller, 2010) are only some reasons that have made scholars, like Zittrain (2008), worry that digital systems may be pushed back to the model of locked-down devices centrally controlled information appliances.

Hence, a battle is emerging among agents (several governments and corporations) that are trying to turn the Internet into a tightly controlled information medium, and user communities that are trying to keep the medium independent (Kostakis, 2013). This battle certainly affects the design processes of smart cities as well, because it has a direct relation with the involved stakeholders.

This chapter attempts to simplify possible outcomes by using two axes or polarities that give rise to four possible scenarios (see Figure 7.1) and then tries to adapt the evolution of the smart city in this context. The chapter concludes by drawing some assumptions about what should determine the ideal selection for a smart city.

The Two Axes and the Four Quadrants

The first axis concerns the polarity of centralized versus distributed control of the infrastructure; the second axis relates to an orientation

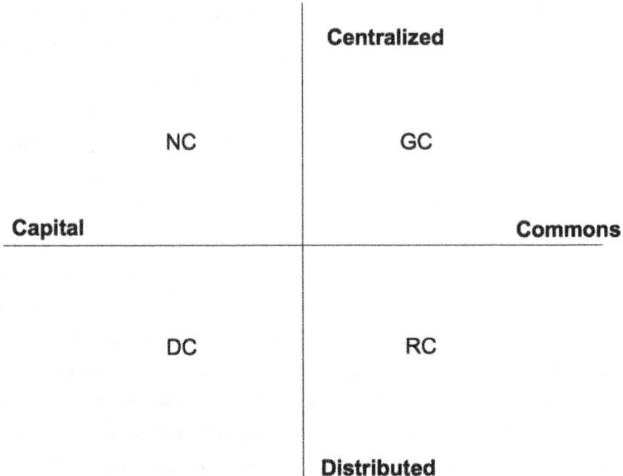

Figure 7.1 The four quadrants of future scenarios.

toward the accumulation or circulation of capital versus an orientation toward the accumulation or circulation of the commons.

First, we introduce the concepts of "netarchical" and "distributed capitalism." Before describing in detail the two forms that shape this emerging model, it is important to highlight their basic difference. Netarchical and distributed capitalism may both be profit oriented, but they are also based on various technological regimes' dependence on the structure of every project's back end. User-oriented technological systems generally have two sides. The front end is the side that users interact with, and the only side visible to them. The back end, however, is the technological underpinning that makes it all possible. This is engineered by the platform owners and is invisible to the user. Hence, a front end that enables a P2P social logic among users can often be highly centralized, controlled, and proprietary on the back end; forming an invisible technosocial system that profoundly influences the behavior of those using the front end, by setting limits on what is possible in terms of human freedom. Then, we present the remaining quadrants, that is, resilient communities and global commons whose ultimate goals are commons driven.

Netarchical Capitalism (NC)

We define "netarchical capitalism" as the first combination (upper left) that matches centralized control of a distributed infrastructure with an orientation toward the accumulation of capital. NC is that fraction of capital that enables and empowers cooperation and P2P dynamics, but through proprietary platforms under central control. While individuals will share through these platforms, they have no control, governance, or ownership over the design and the protocol of these networks/platforms (e.g., Facebook or Google). Typically under conditions of NC, sharers will directly create or share use value, but the monetized exchange value will be realized by the owners of capital. While in the short term it is in the interest of shareholders or owners, this also creates a longer-term value crisis for capital, because the value creators are not rewarded, and have no purchasing power to acquire the goods that are necessary for the functioning of the physical economy.

Distributed Capitalism (DC)

The second combination (bottom left), called "distributed capitalism," matches distributed control but with a remaining focus on capital accumulation. The development of the P2P-driven currency Bitcoin and the Kickstarter crowdfunding platform are representative of these developments. Under this model, P2P infrastructure is designed in such a way as to allow the autonomy and participation of many players, but the main focus rests on profit making. In Bitcoin, all the participating computers can produce the currency, thereby disintermediating large centralized banks. However, the focal point remains on trading and exchange through a currency designed for scarcity, and thus must be obtained through competition. Furthermore, Kickstarter functions as a reverse market with prepaid investment. Under these conditions, any commons is a by-product or an afterthought of the system, and personal motivations are driven by exchange, trade, and profit. Many P2P developments can be seen within this context, striving for a more inclusionary distributed and participative capitalism. Although they can be considered as part of, say, an antisystemic entrepreneurialism directed against the monopolies and predatory intermediaries, they retain the focus on profit making. Distribution, here, not meant locally, though, as the vision is one of a virtual economy, where small players can have a global compact, and create global aggregations of small players.

Resilient Communities (RC)

Distributed control with a focus on the commons is what we call the "resilient communities" (bottom right). The focus here is mostly on the relocalization and re-creation of local community. It is often based on an expectation for a future marked by severe shortages of energy and resources, or in any case, increased scarcity of energy and resources, and takes the form of lifeboat strategies. Initiatives like the Degrowth movement or the Transition Towns, a grassroots network of communities, can be seen in that context. In extreme forms, they are simple lifeboat strategies, aimed at the survival of small communities in the context of generalized chaos. What marks such initiatives is arguably the abandonment of the ambition of scale while the feudalization of territorial integrity is considered mostly inevitable. Even though global cooperation and web presence may exist, the focus remains on the local. Most often, political and social mobilization at scale is seen as not realistic, and doomed to fail. In the context of our profit-making versus commons axis, though, these projects are squarely aimed at generating community value.

Global Commons (GC)

This approach (upper right) is against the aforementioned focus on the local, focusing on the global commons. Advocates and builders of this scenario argue that the commons should be created for and fought for on a transnational global scale. Although production is distributed and therefore facilitated at the local level, the resulting microfactories are considered as essentially networked on a global scale, profiting from the mutualized global cooperation both on the design of the product and on the improvement of the common machinery. Any distributed enterprise is seen in the context of transnational phyles, that is, alliances of ethical enterprises that operate in solidarity around particular knowledge commons. In addition, political and social mobilization, on regional, national, and transnational scale, is seen as part of the struggle for the transformation of institutions. Participating enterprises are vehicles for the commoners to sustain global commons as well as their own livelihoods. This latter scenario does not take social regression as a given and believes in sustainable abundance for all humanity.

Discussion

These four scenarios differ in their vision for the prime focus of the accumulation of value, either for the benefit of global shareholders, for a network of small for-profit enterprises, for the local community, or for transnational commons. It can be argued that the prevalence of each scenario will have different impacts on the smart city model to be adopted.

All four scenarios take the existence of P2P-enabling infrastructure as a given, and mutualize both immaterial and material resources to obtain economies of scope. Indeed, while economies of scale are advantageous in the context of temporal eras dominated by an abundance of resources and energy—that is, producing more of a thing creates competitiveness—economies of scope become essential in periods of increased energy and resource scarcity—that is, doing more with less. Open source is mutualization of immaterial resources such as knowledge, which become operative for the whole of humankind, rather than fragmented and privatized through intellectual property. The mutualization of physical resources increases the efficiency of resource and energy use, and combats the idleness of physical resources and the waste that is inherent in fragmentation.

The new P2P production modalities are global-local (or glocal). While they enable production at the local scale through microfactories using distributed manufacturing technologies, both the knowledge work on the product and on the machinery can be global. As a general rule, one can say that the principle is this: "what is heavy is near, what is light is far"; thus we design global, but manufacture local responding to certain needs. Cooperation on the immaterial productive processes (i.e., design) is maximized, but the global transportation of material good is minimized. This new productive model should be carefully considered during policy making for urban development as it can have a profound impact on the city itself.

In our four scenarios, what differentiates the strategies are first of all, the aim of the cooperation, that is, are they aimed at capital accumulation, or at improving the circulation of the commons? And second, where is the focus of control? Is control distributed through free self-allocation by commoners who can affect the governance and design of their infrastructure of cooperation? Or is the design of the infrastructure in the hands of centralized privately owned platforms? The answer to the these questions will probably define the final form of the so-called "smart city."

If we want to locate the "smart city," as it is conventionally understood, in the context of our scenarios, we should look at the top-left quadrant of netarchical capitalism (see Figure 7.1). What we have for the time being is smart cities in terms of ICT deployment and not actual smart urbanism. Citizens are able to contribute by providing "big data," which are gathered from the utilization of an array of sensors throughout a city, to offer governments/firms solutions to their needs. But as it happens in this scenario, control and governance in today's smart cities are located within a single proprietary hierarchy, where the main motive is profit maximization. As a result, it is questionable whether citizens actually take part in the decision-making process, in order to meet their true needs, or just constitute another source of information without knowledge and influence at the back end.

The circumstances could be slightly different in the distributed capitalism scenario, where control is located in the network of participating for-profit entrepreneurs. Here, citizens may enjoy an increased capacity to influence the shaping of smart city infrastructure, leading to more transparent and democratic decision making for specific issues. However, profit maximization remains the ultimate goal for all parties involved. This can, arguably, have a negative impact on the aforementioned decision-making process and lead to unsustainable outcomes.

The further we move toward the right quadrants, those of resilient communities and global commons, the higher the potential for bottom-up civic engagement and support of citizen empowerment and user-driven innovation. In the local community model, control is located in a particular geographical territory, and depends on the governance model of the initiating community. The adoption of this scenario while planning a smart city—or even a smart town—could lead to successful practices, as designing in a smaller scale includes strong predefined goals that can be bounded with measurable results and quick decision making. Contrary to similar interventions in big cities, a small area means a smaller chance for failure. However, the knowledge and know-how produced in this case may not be widely applicable or even available for adoption elsewhere, due to the fact that it is locally oriented. This potentially hinders the circulation of the commons and the subsequent diffusion of innovation regarding smart cities.

At the grander scale of the global-local commons model, governance is located in the triarchical model of the community practicing the social self-allocation of resources, of the for-benefit associations that manage the physical infrastructure of cooperation (e.g., the multitude of Free/Libre Open Source Software Foundations) and of the entrepreneurial alliance that cooperates around the same commons. In this model, it is essential that the commons orientation is guaranteed by new governance models of the participating entrepreneurs. For example, in the case of the largely corporate Linux Commons, open source code commons are clearly integrated in the processes of capital accumulation of the participating for-profit enterprises. A countermodel would require the creation of commons-friendly, ethical enterprises, consisting of the commoners themselves, who also control their own governance and have ownership. Such enterprises would be legally structured so that theirs is an obligation to support the circulation of the commons. We suggest a plural form of ownership that combines maker ownership (i.e., a revisiting of worker ownership for the P2P age), with user ownership (i.e., a recognition that users of networks co-create value, and eventually a return for the ethical funders that support the enterprise). In this model, profit making is allowed, but profit maximization remains a taboo.

The manifestation of the smart city in this scenario is highlighted by wide citizen engagement while designing and implementing interventions and an ongoing circulation of the commons, which promotes continuous innovation and knowledge diffusion. In this case, the production of commons on a global scale will lead to a more sustainable city model, which could perform better than the current dominant model while solving a number of systemic problems.

To enhance user participation, the creation of a unique culture is vital. This can be accomplished through implementing small-scale, low-cost actions that have little bureaucratic requirements and encourage citizens to reclaim common open spaces in the urban environment. These processes should serve as a user-driven platform for the local community and lead to the creation of a robust paradigm aiming to collaboration.

Toward that direction, governments and local authorities should provide appropriate facilities to enable the deployment of participative ways of working, which will help in producing social innovation outcomes, that is, commons. This could be done by promoting the creation of collaboration spaces, such as microfactories, all over the city and creating wired and wireless networks that will enhance the connectivity between citizens. Moreover, the establishment of social enterprises should be promoted. This will certainly lead to the development of business models, but instead of seeking easy financial gains, social enterprises will be focusing on sustainability and development in the long term.

After ensuring the existence of the basic infrastructure for a commons-driven smart city, the next step would be to integrate them into everyday social interaction and make all the data available to the citizens in a format that they can use. Because several cities will deploy different infrastructure and adopt various approaches, this procedure may become quite challenging. In order for locally produced innovations to diffuse and be adopted globally, the aforementioned infrastructure should comply with some "standards" that will enhance interoperability. These "standards" should be based on open source technologies, so they would be easily accessible, transparent, and open to modification and adaptation to local conditions and individual needs.

Conclusion

One of the most fundamental characteristics of a smart city should be its direct link with the needs and concerns of urban residents. However, it has already been observed that this citizen perspective is often ignored in the smart city discussion. While technology is a powerful tool able to help improve urban infrastructure, citizen engagement is essential to make cities truly sustainable and livable.

In the discussion above, we argue that different applications of certain productive infrastructure have different impacts on urban life, depending on the model of governance and strategies of citizenship they embody. Notwithstanding the fact that community-driven, commons-driven, and distributed versus centralized for-profit–driven infrastructure coexist, smart cities will be organized differently depending on the dominance of any of the four scenarios.

What is needed, in our view, is a more commons-driven smart city that will provide the capacity for open participation and democratic problem-solving practices that can potentially lead to social, environmental, and economic sustainability.

Acknowledgment

V. K. acknowledges support by the grants SF 014006 "Challenges to State Modernization in 21st Century Europe" and ETF 8571 "Web 2.0 and Governance: Institutional and Normative Changes and Challenges."

References

Bauwens, M. (2005). "The political economy of peer production." *CTheory Journal.* Retrieved March 3, 2013, from http://www.ctheory.net/articles.aspx?id=499

Benkler, Y. (2005). *The wealth of networks: How social production transforms markets and freedom.* New Haven, CT: Yale University Press.

Boyle, J. (1997). "Foucault in cyberspace." Retrieved March 3, 2013, from http://www.law.duke.edu/boylesite/foucault.htm

Caragliu, A., Del Bo, C., and Nijkamp, P. (2009). "Smart cities in europe." Series Research Memoranda 0048, VU University Amsterdam, Faculty of Economics, Business Administration and Econometrics.

Kostakis, V. (2013). "At the turning point of the current techno-economic paradigm: Commons-based peer production, desktop manufacturing and the role of civil society in the Perezian framework." *TripleC,* 11(1), pp. 173–190.

Kostakis, V., Fountouklis, M., and Drechsler, W. (2013). "Peer production and desktop manufacturing: The case of the Helix_T wind turbine project." *Science, Technology & Human Values,* 38(6), pp. 773–800.

Mueller, M. (2010). *Networks and states: The global politics of Internet governance.* Cambridge, MA: MIT Press, p. 160.

Zittrain, J. (2008). *The future of the Internet – and how to stop it.* New Haven, CT: Yale University Press.

8
Smart Cities: Toward the Surveillance Society?

Tarun Wadhwa

Information is the lifeblood of the "smart city." While it's the people, ideas, and commerce that shape how a city looks, in many urban areas around the world, it is streams of data and algorithms that will determine how it actually operates. It's not that the movements, scenes, and interactions that have always made up life in a city are changing – it's just that now those can be recorded, understood, and influenced with little human intervention.

As cities increasingly have to compete on a global scale to attract attention, talent, and investment, local governments are turning to technologies to stand out. While the reasoning is somewhat different in each case, leaders in cities around the world are looking to "smart" solutions to improve efficiency, security, and well-being. Information and communication technologies have now advanced in affordability, precision, and utility to the point where they are capable of making a significant difference in how a city can function.

The once-distant fantasy of a city capable of partially running itself is now actually becoming feasible, and creating an urban digital infrastructure is coming to be seen as an imperative, not an experiment. With the already-rapid pace of urbanization expected to increase globally over the next few decades, there are huge challenges ahead that must be addressed. Established technology giants are lining up, ready with solutions to cash in on the need and the hype. The opportunities are massive, but the answers are still elusive.

For all the excitement and promise, the process of making cities "smart" is far more complicated than just adding a technological layer onto an existing society. Besides the usual challenges related to execution and adoption, the prospect of blanketing our cities with sensors raises some major concerns related to privacy, security, and control. In the rush to become "smart," it's questionable whether city officials are really thinking through the cybersecurity implications of embedding computers into vital urban functions. Issues related to how personal information is processed, handled, and shared are still not being given nearly enough attention.

Smart cities further centralize power. The technologies at their core allow various public and private entities deep abilities to track and influence your life on a scale never before possible. While there are certainly benefits to living in a "smart" environment, what's less discussed is what the trade-off actually is. Life in the smart city is life under constant surveillance; whether that is a beneficial or harmful thing often comes down to oversight, awareness, and preparation.

Connecting the resident and the city

Before attempting to determine what impact making a city smart can have on its residents, it is necessary to sort through what making a city smart actually means. It turns out that groups of people can mean very different things when talking about what is "smart." Generally speaking, the overall outcome is to improve residents' quality of life through the use of technology.

Anthony Townsend, a researcher and the author of *Smart Cities: Big Data, Civic Hackers, and the Quest for a New Utopia*, argues that the definition of what makes a city smart depends on whom you ask:

> There are many different visions of what the opportunity is. Ask an IBM engineer and he will tell you about the potential for efficiency and optimization. Ask an app developer and she will paint a vision of novel social interactions and experiences in public places. Ask a mayor and it's all about participation and democracy. In truth, smart cities should strive for all of these things. (2013)

Smart cities are created to help address needs, but in an environment as complex as a city, those needs can vary quite a lot among different populations.

More specifically, according to Forrester Research, there are seven critical city infrastructure components and services that technologies can be used to improve: city administration, education, health care, public safety, real estate, transportation, and utilities (Washburn & Sindhu, 2010). The group describes the technologies involved as "integrated hardware, software, and network technologies that provide IT systems with real-time awareness of the real world and advanced analytics to help people make more intelligent business decisions." The data collected by these systems come from a variety of public, private, and personal currently existing and freshly installed sources: surveillance cameras, connected sensors, equipment, phones, navigation systems, monitoring stations, meters, social networks, and information created through crowdsourcing (Ferrero, 2013).

One government report estimates that by the end of the decade, the entire market could be worth over $400 billion (Smart City Market, 2013). Yet the "killer applications" for smart technologies are still being discovered; it is still quite early to estimate what type of economic impact the industry might make. There is still a healthy degree of skepticism as to whether these technologies can actually improve the way a city functions, or whether they just improve the facade.

In short, cities have extremely limited resources and it's still not clear what functions of a smart city are "worth" the investment; however, that will change as the industry grows. Many of the problems that cities face, from traffic to air pollution, are the same, and if a method is truly an improvement, you can expect it to spread widely. The question isn't whether cities will become smart; it's which technologies and approaches actually create value for all stakeholders. Those that are truly "smart" will no longer be known as that; they will follow the path of other successful technologies and just become the default expectation.

How smart is too smart?

From a policy perspective, smart cities are fascinating laboratories filled with promise and peril. They are living, breathing, vibrant test cases showcasing the most pressing technological and social issues of our time. Residents of a smart city are essentially a captive audience; the things they see, do, and experience every day are the variables that are modified. Just like users of the largest sites on the Internet today have little idea what is going on behind the scenes, so too residents of a smart city will only be able to see what is immediately revealed to them.

The dizzying pace of technological advancement and the aggressive nature of technology companies has already made it extremely difficult to isolate the changes that technologies are making to our lives – smart cities are the same, but on a larger scale. It will be virtually impossible for a resident of an advanced smart city to know what is being collected and determined about them at any time. Awareness and outrage over abuses are what reins in some of the worst practices of any industry; yet in a smart city, it is far less likely that it is an influential counterweight.

In Europe, the Article 29 Data Protection Working Party, an independent government advisory body comprise of representatives from each nation working on security issues, recently adopted a guidance for how European Union (EU) legal framework should work with the security challenges posed by Internet of Things technologies. Their largest concerns were over lack of controlling who is able to access information, issues of improper consent, detection of behavior patterns and surveillance, and increased security vulnerabilities. Their solution was to include device manufacturers, platforms, application developers, and data platforms as responsible stakeholders that have to fulfill obligations. While this may be too onerous for many nations, responsibility for security certainly needs to be expanded (Kuschewsky, 2014).

People everywhere have been shown to have extremely conflicting attitudes about privacy and sharing. Oftentimes, they will state one set of beliefs, then go to completely violate them in their actions. Other times they will find a particular example of technology being used in an invasive way as offensive, but they will accept large-scale systems of dragnet government and corporation surveillance as the norm. Yet as problematic as this is, it has allowed consumers to draw certain lines around the ways their information can be used and they can be treated. It's unclear how that can happen in a smart city environment, unless there is a particular focus on enforced information disclosure, which has proven to be quite rare.

Take, for example, the attempt of a United Kingdom–based company, Renew, to test a network of Internet-enabled recycling bins in London before the 2012 Olympics. Unknown to the hundreds of thousands of pedestrians, some of these bins were collecting personal information from their phones – recording the unique MAC addresses for the purpose of serving them up personalized ads later. Once the plan was revealed the company was forced to shut down the trial, but the proof of concept showed just how easily a company could collect your personal information for whatever purpose they decided (Shubber, 2013). There are dozens of examples of situations like this from around the world, each a loud collective statement of what is permissible behavior.

A trash bin can be easily removed, but there is now also such a thing as Internet-connected pavement which will not be so simple to remove (Jones, 2012). Add this to the list of streetlights, billboards, and signs that are capable of measuring and analyzing you. So many products these days are coming with Internet-connected capabilities for no reason in particular. A smart city dweller has to live under the assumption that they are constantly having their personal information collected by everyday items, without any idea of whom else it will be shared with. Objects that we would never normally consider to be invasive are now part of a larger surveillance apparatus, and they are so ubiquitous that it is not possible to functionally avoid them.

When every move is measured

Travelers visiting Newark airport in the state of New Jersey probably wouldn't be surprised to hear that they are being watched, but they may be surprised by what is keeping an eye on them. A Silicon Valley–based company called Sensity is selling what it calls "smart lights" to be installed at parking garages across the nation (CBS News, 2014). The technology is presented as an innovation in efficiency, but the lights do a lot more than that. They come included with a sophisticated array of sensors and cameras that "monitor security and the flow of foot traffic." With LED lights showing themselves to be 90% more efficient than their predecessors, the company is using the language of environmentalism to justify an expansion of surveillance.

There is a paradigm shift coming to the world of surveillance, something that will change the way we look at the cameras that look at us. The "pact" we have made as a society with technology is rooted in the assumption that nothing is done with the vast majority of information collected on us. Part of the reason, in the developed world, people have become so accepting of surveillance cameras has been because when footage is taken of us walking through a public or private place, the vast majority of the time nobody will ever look at it. Even if our e-mails are being collected on government servers, there is the belief that nobody will really take the time to actually read it.

Thanks to advances in data processing, machine learning, and computer vision, we are nearing a world where the surveillance cameras that currently record our every move are also able to analyze them – to try to understand our emotions, behaviors, and intentions. These technologies are in their early stages of development, but they are still deployed widely in public places, private companies, and large events around the world for everything from finding a bomb to spotting a potential shoplifter (Wadhwa, 2012). Their usage is only growing as their accuracy and utility increases, and although we are still a few years away from their widespread usage, these changes must be considered when we determine what role cameras should have in our lives.

At some point, adding cameras and sensors to existing fixtures to collect new types of data becomes trivial. They will soon just be a matter of installing cheap parts and updating software. Advancements in the size, sensitivity, and design of cameras are driving down costs and opening up a world of surveillance possibilities. An interesting parallel here is the widespread adoption of Automated License Plate Recognition by police forces across the developed world. With little scrutiny or oversight, advances in cameras and character recognition have allowed law enforcement to take advantage of a lax regulatory environment to build a massive surveillance system that is quickly making it very difficult to drive anywhere without authorities having a record of it.

In order to have a productive conversation about smart cities, we need to understand the nature of technology – the things we use it for change based on our needs and intentions. Allowing our every movement to be tracked requires putting a great deal of faith in the future: that the information will stay secure, that it won't be misused, and that it's value will be respected by those in charge. The approach that cities sometimes take of installing a new technology then disclosing what they will use it for after will only serve to erode trust overtime (Fisher, 2014). Residents must understand the power of the technologies they are entrusting their leaders with, and leaders and corporate partners must acknowledge it.

Securing a city of vulnerabilities

Although they are often separated in value judgments, privacy and security go hand in hand. When malicious actors are able to steal or access your personal information, there can be no real privacy. And unfortunately, to a hacker, the smart city is a playground unlike any other.

Reflecting upon plans for India to build 100 smart cities in the near future, Sanjay Rohatgi, president at security giant Symantec India, sees the enormity of the problem – "with increased data generation within the city infrastructure, the smart city soon becomes a tempting proposition for cyber-criminals because of its technological diversity and sophistication." While laying out the potential benefits for cities – from increased security to better urban planning – Rohatgi argues that security needs to be part of a smart city's blueprint. In order to prepare the country for the transition over to a sensor-filled urban environment, Rohatgi recommends several proactive measures: establishing a governance framework, responsible compliance regulations, robust authentication of users, balancing security considerations with convenience for cloud computing, paying special attention to critical infrastructure, and building out the policies to support the operations (Rohatgi, 2014).

The problems of security in the smart city are largely a magnified extension of the problems of the Internet of Things and Internet-connected devices in general. Security still isn't taken seriously enough, and with a lot of these systems, there isn't a way to easily fix and patch things up (Wadhwa, 2014). The recently uncovered enormous Heartbleed vulnerability is present in many embedded devices that are widely used for a variety of purposes (Cantrell, 2014). It is likely that many of these will never be patched. The prospect of another zero-day vulnerability looms large; it's a matter of when, not if.

Technical developers know that user experience, especially as it relates to security, is a trade-off of time and constraints. There will almost always be more bugs than can be fixed, more problems than there are people to deal with it. Entrepreneurs understand this and know they have to prioritize. The best software available today is still vulnerable and flawed. The difference is, a computer or mobile application is contained in a small environment, but in a smart city, the problems play out on a massive scale.

Even researcher and author Townsend (2013), a longtime proponent of smart cities, has come to question whether we are walking into a computer error–filled nightmare. He argues:

> Smart cities are almost guaranteed to be chock full of bugs, from smart toilets and faucets that won't operate to public screens sporting Microsoft's ominous Blue Screen of Death. But even when their code is clean, the innards of smart cities will be so complex that so-called normal accidents will be inevitable. The only questions will be when smart cities fail, and how much damage they cause when they crash.

He even calls the smart city "as brittle an infrastructure as we've ever had."

But even more than the potential for errors is the lack of preparation for the consequences. The largest concern might be how little we actually know about what vulnerabilities lie ahead:

> The pervasiveness of bugs in smart cities is disconcerting. We don't yet have a clear grasp of where the biggest risks lie, when and how they will cause systems to fail, or what the chain-reaction consequences will be. Who is responsible when a smart city crashes? And how will citizens help debug the city? Today, we routinely send anonymous

> bug reports to software companies when our desktop crashes. Is this a model that's portable to the world of embedded and ubiquitous computing? (Townsend, 2013)

The fact that these questions don't have clear answers means that the corporations and consultants pushing smart city solutions as a panacea should take a hard look at the negative potential impact of what a rapid adoption might lead to (Townsend, 2013).

In terms of technical improvements to the smart city landscape, experts have called for more secure data processing and storage, improved access control, smarter data aggregation techniques, standardization efforts, secure monitoring, and privacy-respecting context-aware computing (NetWorks, 2011). Some of these will surely be achieved. Whether they will happen fast enough – and be implemented properly – is unclear as of yet.

Building on top of a broken foundation

The greater problem with security in smart cities is that it is so reliant on the performance of underlying technological systems. If there are interruptions in cloud services, problems with power distribution, or even a system like GPS has a critical error, all of the functions of the city are at risk. This problem is not unique to these systems. Developers can only do so much; they do not control the foundation they are building upon.

It's common knowledge that the Internet was not built to be secure; however, it's still quite rare to see individuals and companies let this fact impact their decision making about what to connect and share. Daniel Suarez, a sci-fi writer who has spent many years working in cybersecurity, explained to *Forbes* that we need to create something better:

> What we need is an Apollo-like national project to build a new, secure network for critical infrastructure that would use a separate protocol, proprietary hardware, dedicated fiber-optic lines and powerful encryption to eliminate all but the most elite interlopers. This wouldn't replace the Internet; it would only be used where identity and trust are critical. (Hill, 2014)

A long-term solution to these problems absolutely requires the messy, unpopular process of addressing the structural issues (Hill, 2014).

One of the largest concerns for any government is the securing of their energy supply and power grid. Recently, much attention has been given to the fact that many of our most important infrastructure systems run on technologies that are decades old. At Black Hat Europe, in October 2014, researchers Javier Vazquez Vidal and Alberto Garcia Illera demonstrated how they were able to reverse engineer smart meters, finding blatant vulnerabilities in commonly installed technologies. They discovered that each meter tested contained the same encryption keys. If a hacker has access to the key, he/she is able to remotely power off systems (Higgins, 2014).

This is just a preview of what select researchers have chosen to disclose. As far as what state actors and organized crime can do is far beyond this. Security researchers recently found a sophisticated piece of malware nicknamed Dragonfly – thought to be of the sophistication of the Stuxnet virus – that is specifically designed to target industrial control systems managing electrical, water, oil, and other critical systems (BBC, 2014). It is thought the virus was created for espionage purposes by hackers in Eastern Europe.

While it is true that there have been no catastrophic smart city hacks causing major, lasting damage yet, we don't need to wait for one to happen to prepare for the fallout. If we truly want a completely connected lifestyle, anything less than revamping the entire foundation would be insufficient. The status quo of cybersecurity is in poor shape, which is why there is a new major data breach almost every week. More and more countries are developing offensive cybersecurity capabilities, and the systems running our critical infrastructure are becoming even more obsolete every day.

Therefore, it would be wise to keep certain functions just as they are without fully understanding the vulnerabilities that making them "smart" would entail. We are still many years away from a foundation that can be trusted – and we will surely make improvements as we entrust more and more of our lives to networks. In the absence of something that is truly proven to be reliable, the status quo may actually be the better option.

Privacy cannot be an afterthought

The stakes are too high for privacy to be on the bottom of the list of considerations; the early decisions made in how smart cities will operate have enormous ramifications. Technologists, city officials, and residents all must take a proactive approach to keeping personal information safe. Yet that is easier said than done when there is so much information being collected.

In the 1990s, Ann Cavoukian, Information and Privacy Commissioner of Ontario, coined the term "privacy by design" as a methodology for how to integrate privacy into a product. The process focuses on "embedding privacy into the design, operation, and management of information technologies and systems, across the entire information life cycle." Although it was developed over a decade ago, the principles behind it have only become more useful and applicable.

Cavoukian sees a lot of potential to apply privacy by design to how we organize smart cities. The problem comes down to how information is shared, she explains; extra attention is needed "when there is the possibility of unauthorized services or third parties discovering personal information, such as individuals' personal habits, behaviors, and lifestyles, and using this information without their consent for secondary purposes, like marketing."

She cites the smart grid in Canada and California as an example of success. So much information about a person's lifestyle can be determined from when and how he/she uses electricity. Because there was awareness of the sensitivity of this information beforehand, steps were taken to mitigate potential privacy harms.

If residents do not understand what they are trading, they will be a lot less inclined to want to participate in future programs. Companies shouldn't take note of this just because it's the right thing to do, but because it's good business and good for the industry. Wim Elfrink, who heads up Cisco's smart city team, issued a warning to others to approach this issue with appropriate seriousness, stating that "having security policies, having privacy policies is a given. I think you have to first give the citizens the right to opt-in or opt-out" (Datoo, 2014).

The notion of being able to opt out of data collection in a smart city is questionable. While that can be helpful in the early stages of a new application of a technology, at some point there is really no choice – a sensor is going to record your movements, a camera is going to monitor your steps. The only way to "opt out" is to leave the city, which is not a realistic option. There is a small window to get these issues right, after which it becomes far more difficult.

You can't be anonymous in public

While huge amounts of our personal information is collected every day, we are constantly told that it is not a problem because it's not actually personal information – the data is "anonymized." The idea is that information that could lead to discovering a person's identity is removed, therefore they should not object to pervasive monitoring. Yet study after study is proving this to be not quite true.

According to reidentification researchers, it takes only a few bits of information to determine the identity of a person (Narayanan & Shmatikov, 2010). That could be significantly less depending on what type of information we are talking about. Knowing a person's browser type or gender is different from knowing their zip code or last name. That is even more true when it comes to behavioral information – with the amount of personal information available on the Internet today, it can at times be quite straightforward to reidentify a person from publicly released data sets. The frequently made promise of anonymizing data is oftentimes a false one.

There should be much more attention placed on making sure data are properly handled. Internal and external audits, regular reports, and laws requiring notification of certain issues are all productive steps to this end. While this practice is still better than its absence, it can't just be taken for granted. Companies need to take these risks far more seriously, paying a great deal of attention to the quality of their processes (Meyer, 2013). Residents would be better served questioning what information is being collected, how it is being shared, and what can be done with it.

Technological problems and policy solutions

The technological limits to surveillance are continuing to disappear. The revelations of former National Security Agency (NSA) contractor Edward Snowden revealed how the U.S. government, along with several others, is conducting massive, sweeping dragnet surveillance on huge populations across the world. Just years ago this type of collection would be unfathomable, but with the price of data storage dropping rapidly, and the processing power available to do deep analytics increasing exponentially, new types of surveillance keep becoming not only possible but also economical.

Privacy researcher Ashkan Soltani, in a landmark study for *The Yale Law Journal*, found that advances in technology were decreasing the cost of surveillance by magnitudes, sometimes even greater than that (Soltani & Bankston, 2014). Based on recent Supreme Court cases involving tracking an individual's location, Soltani found that the difference between following somebody with a car and tracking a cell phone signal over a 28-day period is almost 300 times less. In a smart city environment, where there are a countless number of sensors collecting information, the cost of monitoring a person's every move becomes negligible.

Like many issues related to policy, proper training and planning on a day-to-day basis can impact how often abuses occur. Addressing the vulnerability of the system can lead to building accountability into regular processes. A well-maintained system is harder for one party or group to take advantage of.

There are so many different databases owned by so many different actors, each with its own priorities and constraints. Therefore, there needs to be other safeguards in place – and this is where policy approaches are not just helpful, but absolutely necessary. Laws, standards, and their rigorous enforcement are perhaps one of the only things that can keep a smart city from becoming a one-sided power grab.

Nothing less than comprehensive approach will do

Smart city security issues present an unusually difficult challenge for regulators. There is no consistent policy framework, the field is too early for best practices, and the technology companies that install these systems follow local law. Legislation would need to be passed at the national level to be completely effective. Additionally, outside of special cases protecting critical infrastructure, there is little history of attempts to pass legislation to force greater consideration of what happens when things go wrong. Related regulation is usually a derivative of national privacy and data ownership laws, which vary greatly in strength across the world.

In an article for *IEEE Communications Magazine*, a group of academics suggests adopting "Privacy Enhancing Technologies" used in privacy models for databases and location-based services, to build a new model for "Citizens Privacy." By borrowing from another field of computing, they are able to draw lessons from something that is already in place. Instead of testing a new strategy in the wild, it makes sense to reflect on what has happened with databases over the last few decades. The authors split database privacy issues into three dimensions related to the main actors involved: reidentification of respondents, hiding the query of users, and sharing of information between database owners.

They propose a "five-dimension model for citizens' privacy in smart cities. The identified dimensions are: identity privacy (protecting who you are), query privacy (protecting what you are looking for), location privacy (protecting where you are), footprint privacy (protecting what you've done), and owner privacy (protecting what is shared). It reflects a belief that a citizen should be able to access services without telling everyone who they are and what they are trying to do" (Martinez-Balleste, Perez-Martinez, & Solanas, 2013).

Smart cities are too encompassing of many different subjects just to be dealt with piecemeal. We should press governments and corporations to define their limits, just as residents should define theirs. While it may be too early, at some point this will need to be clarified; we will need a "resident's bill of rights" for life in the smart city.

Sensors or censors?

The company that knows what time you go to bed, what medicines you take, and when you are at your most vulnerable state can do a lot to improve your quality of life. It can help you make sure you have the things you need, and give you reminders to take care of your health. Or it can use that information to serve you advertisements and raise your insurance premiums. If history is any indication, it is likely that both things will happen.

The nature of smart cities is the same as a totalitarian society in many ways – everything is known to central authorities. The idea is that this information will only be used to improve your life, but the reality is probably something in between. This shouldn't be taken lightly or dismissed; the use of ubiquitous tracking technology creates a major imbalance in power.

There is a concern that all this technology will create a chilling effect on residents. When they don't know who is watching or what they are doing, they would be far less inclined to discuss potentially sensitive or taboo subjects, perhaps to the overall detriment of society. Although privacy is often trivialized as something only for those with secrets, many of us have information, thoughts, and feelings we would not want publicly disclosed.

It should become a standard part of every smart city installation to specify exactly what information is collected and what can be done with it. The best protections for residents would be robust controls over who can access the data streams and how long they will be retained for. Generally speaking, the less time personal information is held and processed, the greater the chance that rights will be respected. An even better approach, that few in the technology industry have been willing to even consider, would be to let consumers decide who has access to their data – perhaps even with the option of selling specific information sets to parties that may be interested.

Yet even these types of controls can't change the pervasive nature of smart city surveillance. For these technologies to truly have a completely positive impact, there would need to be a benevolent, enlightened government that is not primarily interested in the preservation of power. While this is certainly possible, it's far from the norm, and unrealistic to expect in every country.

Technology doesn't change priorities

Who does the smart city serve? Are all residents treated exactly equally, or does it continue the current divisions of society that currently exist? Does a network of computers understand why we have a "bad part of town," and how it got to be that way?

Authors of a recent 10-year forecast exploring inclusiveness in smart cities by Institute for the Future (2011), commissioned by the Rockefeller Foundation, warn that "without this catalyst for cooperation, we may repeat the devastating urban conflicts of the 20th century that pitted central planners like Robert Moses against community activists like Jane Jacobs." How smart city systems are controlled, how they're passed down through changes in government, and who they are "targeted" at will be crucial in determining their impact.

Typically, the most cutting-edge surveillance technologies are used on marginalized populations before they become accepted by society at large. Writer Virginia Eubanks (2014) points out that we aren't surveilled as individuals, but rather as groups of people based on our attributes. Technology simply replicates the existing tensions today, "[b]ecause of the persistence of segregation in our offline and online lives, algorithms and search strings that filter big data looking for patterns, that begin as neutral code, nevertheless end up producing race, class, and gender-specific results." In the end, marginalized communities are singled out for most aggressive uses of scrutiny.

It's dangerous and naive to think that the addition of technology alone will change the priorities of a people. When there is a news story about a city adopting smart technologies, it is usually described as a universally positive thing – the city and its leaders are celebrated for their forward-thinking ways, and few bother to question motives and intentions. But the reality is that some of the cities with the largest needs and the largest opportunities are in countries with questionable human rights records. Smart technologies will help a government achieve whatever it is going to do anyway – and that is not always for the betterment of everybody.

Between peril and potential

Smart cities have a lot of potential to improve the way people live their lives, but their value is far from proven. A study by Intel showed quite different attitudes around the world. Outside of the United States, there seemed to be a far larger appetite for deploying these technologies in the very near future, whereas in the United States, less than 50% believe that citywide data collection was a good idea, with far greater concerns about privacy (Roush, 2014). It's a good thing that the bar is being set high for smart city vendors to deliver on their promises.

The future of the smart city depends greatly on how security and privacy are understood. Unlike an abstract privacy threat or a scary-sounding yet unfamiliar piece of malware, residents won't be quick to forget it if the infrastructure of their city becomes hacked or stops working properly. The improvements that a smart city offers won't be worth it unless it is perceived as trustworthy and reliable. A few bad headlines can do lasting damage to the future prospects of the industry, which should give involved players even more incentive to aggressively self-regulate.

Whether smart cities themselves are surveillance societies is not the question. It's not up for debate whether people who live in these places will be tracked on a large scale. There's no disagreement that a resident's every move in public is subject to be recorded by a variety of technologies. Living in a smart city means existing in a state of normalized surveillance. Whether that is a negative or positive thing, whether that information is used to enrich residents' lives – or to control them – that is what is still to be determined.

References

BBC News (2014, July 1). "Energy firms hacked by 'cyber-espionage group Dragonfly.'" Retrieved from http://www.bbc.com/news/technology-28106478.

Cantrell, T. (2014, April 30). "Heartbleed and its impact on embedded security." Retrievedfromhttp://www.embedded.com/design/safety-and-security/4430090/Heartbleed-and-its-impact-on-embedded-security.

CBS News. (2014, June 30). "Technology in LED smart lights raises privacy concerns." Retrieved from http://www.cbsnews.com/news/technology-in-led-smart-lights-raises-privacy-concerns/.

Datoo, S. (2014, April 4). "Smart cities: Are you willing to trade privacy for efficiency?" Retrieved from http://www.theguardian.com/news/2014/apr/04/if-smart-cities-dont-think-about-privacy-citizens-will-refuse-to-accept-change-says-cisco-chief.

Eubanks, V. (2014, January 15). "Want to predict the future of surveillance? Ask poor communities." *The American Prospect* (January 15), Retrieved from prospect.org/article/want-predict-future-surveillance-ask-poor-communities.

Ferrero, F. (2013, September 20). "Privacy in the era of smart cities – business value exchange." Retrieved from http://businessvalueexchange.com/2013/09/20/privacy-era-smart-cities/.

Fisher, D. (2014, May 19). "Auckland's CCTV plan to watch you." Retrieved from http://www.nzherald.co.nz/nz/news/article.cfm?c_id=1&objectid=11257325.

Higgins, K. (2014, October 1). "Smart meter hack shuts off the lights." Retrieved from http://www.darkreading.com/perimeter/smart-meter-hack-shuts-off-the-lights/d/d-id/1316242.

Hill, K. (2014, June 18). "10 ways to 'fix' cybersecurity." *Forbes*. Retrieved from http://www.forbes.com/sites/kashmirhill/2014/06/18/10-ways-to-fix-cybersecurity/.

Institute for the Future (2011). *The future of cities, information, and inclusion*. Retrieved from http://www.iftf.org/our-work/global-landscape/human-settlement/the-future-of-cities-information-and-inclusion/.

Jones, O. (2012, June 21). "Smart pavement: Coming to a city near you." *IdeaFeed / Big Think*. Retrieved from http://bigthink.com/ideafeed/smart-pavement-coming-to-a-city-near-you.

Kuschewsky, M. (2014, September 29). "Internet of Things poses a number of significant data protection challenges, say EU watchdogs." Retrieved from http://www.natlawreview.com/article/internet-things-poses-number-significant-data-protection-challenges-say-eu-watchdogs.

Martinez-Balleste, A., Perez-Martinez, P., and Solanas, A. (2013, June). "The pursuit of citizens' privacy: A privacy-aware smart city is possible." Retrieved from http://ieeexplore.ieee.org/xpl/articleDetails.jsp?tp=&arnumber=6525606&url=http%3A%2F%2Fieeexplore.ieee.org%2Fiel7%2F35%2F6525582%2F06525606.pdf%3Farnumber%3D6525606.

Meyer, D. (2013, August 27). "This man thinks big data and privacy can co-exist, and here's his plan." *Tech News and Analysis*. Retrieved from http://gigaom.com/2013/08/27/this-man-thinks-big-data-and-privacy-can-co-exist-and-heres-his-plan/.

Narayanan, A., and Shmatikov, V. (2010). "Myths and fallacies of "personally identifiable information." Retrieved from University of Texas website: http://www.cs.utexas.edu/users/shmat/shmat_cacm10.pdf.

NetWorks. (2011). "Smart cities applications and requirements." Retrieved from www.networks-etp.eu/fileadmin/user_upload/Publications/Position_White_Papers/White_Paper_Smart_Cities_Applications.pdf.

Rohatgi, S. (2014, January 3). "Big data & smart cities." Retrieved from http://www.businessworld.in/news/business/it/big-data-&-smart-cities/1721151/page-1.html.

Roush, W. (2014, February 19). "Intel survey: Americans skeptical, yet hopeful, about smart cities." *Xconomy*. Retrieved from http://www.xconomy.com/san-francisco/2014/02/19/intel-survey-americans-skeptical-yet-hopeful-about-smart-cities/.

Shubber, K. (2013, August 9). "Tracking devices hidden in London's recycling bins are stalking your smartphone." *Wired UK*. Retrieved from http://www.wired.co.uk/news/archive/2013-08/09/recycling-bins-are-watching-you.

Soltani, A., and Bankston, K. (2014). *Tiny constables and the cost of surveillance: Making cents out of* United States v. Jones. Retrieved from *The Yale Law Journal* website: http://www.yalelawjournal.org/forum/tiny-constables-and-the-cost-of-surveillance-making-cents-out-of-united-states-v-jones.

Smart city market. (2013). Retrieved from https://www.gov.uk/government/publications/smart-city-market-uk-opportunities.

Townsend, A. (2013, October 7). "Smart cities: Buggy and brittle" [Essay]. Retrieved from http://places.designobserver.com/feature/smart-cities-buggy-and-brittle/38111/.

United Nations Population Fund. (2007, May). "Urbanization: A majority in cities: Population & development." Retrieved from http://www.unfpa.org/pds/urbanization.htm.

Wadhwa, T. (2012, August 30). "The next privacy battle: Cameras that judge your every move." *Forbes*. Retrieved from http://www.forbes.com/sites/singularity/2012/08/30/dear-republicans-beware-big-brother-is-watching-you/.

Wadhwa, T. (2014, January 23). "Do you trust Internet-connected appliances enough to let them run your home?" *Forbes*. Retrieved from http://www.forbes.com/sites/tarunwadhwa/2014/01/23/do-you-trust-internet-connected-appliances-enough-to-let-them-run-your-home/.

Washburn, D., and Sindhu, U. (2010, February 11). "Helping CIOs understand 'smart city' initiatives." Retrieved from http://public.dhe.ibm.com/partnerworld/pub/smb/smarterplanet/forr_help_cios_und_smart_city_initiatives.pdf.

9

Surviving the Electronic Panopticon: New Lessons in Democracy, Surveillance, and Community in Young Adult Fiction

Kerry Mallan

Cities continue to be a subject of interest for researchers working across diverse disciplines. While empirical studies provide insights into people's perceptions of the economic, political, and social dimensions of the urban condition, literature studies also provide a way for writers and readers to reimagine cities and understand their symbolic and cultural significance. When the target audience is young people, there are invariably new lessons that these fictions impart with respect to ways cities as utopian or dystopian spaces may both protect and restrict citizens' rights and freedom. This paradoxical situation of protection and restriction is an outcome of the global use of systems of smart security, including smart sensing tools. While these systems and tools provide information for crime detection and protection of vulnerable citizens, they also amass data on individual citizens, which contravenes their rights to privacy. This much-debated issue of privacy is not only one that is part of legal and social rights discourses but it is also one that is featured in fiction written for young adults: a genre that often taps into young people's developing sense of identity, social entrepreneurism, and political awareness. As children's literature scholars Bradford et al. (2008: 2) note, the field "is marked by a pervasive commitment to social practice, and particularly to representing or interrogating those social practices deemed worthy of preservation, cultivation, or augmentation, and those deemed to be in need of reconceiving or discarding." This chapter examines how a selection of popular young adult (YA) fiction represents and imaginatively constructs cities and their social practices of security in an age of increasing surveillance due to a growing unease about terrorism and other crimes that are often perpetrated within the urban space. The discussion reworks Michel Foucault's concept of the panopticon as a metaphor of surveillance.

In a post–9/11 surveillance culture, the literature for young people testifies to a new set of anxieties that many children and young adults experience. In their study into Australian adolescent students' anxiety about war and terrorism, Summers and Winefield (2009: 178) found that 90.2 percent of the students (mean age 15.2 years) reported "that they had at least some feelings of anxiety for either self, close family/friends, or others" and a need for discussion/expression about issues of war and terrorism. YA fiction can be a stimulus to such discussion as texts often reflect similar anxieties and concerns that many young people experience. The texts may also provide a means for reflecting on the complex social issues that are part of living in technologically advanced societies. In both dystopian and utopian fiction, the city shaped by technology has been a recurring feature of the literature for young people. The city in most dystopian fiction is projected as a negative entity, often an alienating space or a space of danger. However, even within dystopian scenarios there is often a pervasive commitment to a more positive humanistic view with respect to community and belonging. This utopian possibility for a better world or at least to find within the world a place of belonging is characteristic of the field and its optimism for its implied readership.

The YA fiction selected for close analysis includes *The Hunger Games* by Suzanne Collins (2010), *Little Brother* by Cory Doctorow (2008), *Omega Place* by Graham Marks (2007), and the picture book *The Lost Thing* by Shaun Tan (2000). In *The Hunger Games* young people are forced to fight to the death in an arena that is subject to high-technological surveillance and videoing for public consumption and entertainment and recalls Debord's (1988:7) comment that "society has officially declared itself to be spectacular." Panem uses state-of-the-art technology to keep the contestants under the panoptic gaze of the Gamemakers and the general populace, but the twelve districts that surround the capitol experience an uneven distribution of technology, with the poorest living an impoverished agrarian existence. The smart cities represented in *Little Brother* and *Omega Place* are contemporaneous with those that many of its readers experience where electronically ticketed transport systems, mobile technologies, and electronic surveillance are part of the fabric of everyday life. *Little Brother* takes up the discourse of vulnerability that has slipped into American consciousness since the attacks of September 11, 2001, and resulted in increased security precautions and anxieties. Its premise is that citizens must know how to use technology for their own purposes, otherwise they risk losing their democratic rights and freedom, especially when governments can no longer be trusted. The panoptic presence of CCTV and its potential to infringe individual privacy drives the plot of *Omega Place,* where a group of mainly young activists attempt to destroy the cameras in the big cities in parts of England.

The *Lost Thing* offers a different vision of the city. This is not a smart city of the present or future, but a city in an age of entropy. The images of rusty monuments of modernity – the towering, rust-stained concrete structures, the stone statue of a figure with a television head and an outstretched arm pointing somewhere into the distance, the empty highway – stand as stark reminders of a distant past and its long-gone vision of progress and energy. There are no smart transport systems, only the slow movement of trams carrying lifeless commuters. However, entropy is also a term that is used in cybernetics to signify an inevitable loss of information, an unimaginable scenario for our current world with its fetishizing of information. *The Lost Thing* serves as a counter to the other texts in that the panoptic gaze to which the citizens of *The Hunger Games, Omega Place,* and *Little Brother* are subjected is replaced by a social blindness that refuses to see the individual and his/her plight in a world that has lost interest in people and community and the energy that characterizes life in a big city has dissipated. Before discussing the texts I want to locate the central concerns of this chapter within wider theoretical and social contexts with respect to the challenges that an ambient intelligence era poses with respect to privacy.

Every move you make...I'll be watching you

The other side of utopia is dystopia and this duality is no more apparent than in "technotopias" or smart cities. Fast-evolving electronic and computational technologies are central to smart cities and according to Aarts and Ruyter (2009) are responsible for both creating and destroying urban communities. While information and communication technologies will continue to change the use of space in cities and patterns of work, living, and recreation, they will also continue to provide challenges with respect to individual rights, freedom of movement, and privacy. One of the ways in which smart urban initiatives designed for one purpose are appropriated for other unrelated purposes is through function creep. Jeffrey Rosen (2003: 305) describes how urban surveillance measures in Britain originally intended to stop terrorist attacks have given way to function creep: "The cameras are designed not to produce arrests but to make people feel that they are being watched at all times. Instead of keeping terrorists off planes, biometric surveillance is being used to keep punks out of shopping malls."

Innovative traffic systems, such as computerized tolls and high-resolution cameras, designed to reduce congestion, make cities cleaner, safer, and easier for pedestrians and cyclists are also a form of surveillance. As Price (2003: 37) rightly observes, "When these systems are enlisted by law enforcement, however, they may end up undermining the very freedom they seek to encourage, with insufficient mitigating improvements in public safety." Price (2003: 38) reports that while polls taken after 9/11 indicated that the majority of citizens in the United States have become increasingly accepting of surveillance, "justifying the use of traffic systems for law enforcement rests on the assumption that the police tactics work." He cites instances where images captured on cameras set up in cities to scan pedestrian traffic have been wrongly matched with those in a database of known criminals, causing innocent people to be treated as criminals.

Citizens, however, are not always compliant or lacking in agency and can draw on strategies to take control of their place, space, and bodies. The presence of surveillance cameras has promoted groups such as the Surveillance Camera Players to comically adapt and perform plays in front of them (Price 2003: 37). Cameras also record people deliberately showing their faces and demonstrating their disapproval with obscene gestures. These acts of defiance illustrate the capacity of people to resist being docile, impotent, or compliant bodies; they also raise the question of what motivates such actions. By thinking in terms of active bodies interacting with spaces, we can consider motivation not so much as deriving from some moral or immoral purpose, but from a complex set of histories, circumstances, and spatial trajectories, in other words, those influences that prepare us for the choices we make.

Foucault considered subjects as embodied beings capable of acting upon the world and having ability to influence and transform other forces. His view of power at the "extremities" (Foucault 1980) is different from the notion of legitimated or sanctioned forms of power and authority held by the state as a central locus. This idea of power at the extremities is given a material reality in the actions of the Surveillance Camera Players, and even more so in the actions of "smart mobs" or the more playful "flash mobs" (Molnár 2013) that use mobile and smartphones, social networking sites, and blogs as an effective means to harness support and action to effect political protest or prankster acts, making interventions in the urban space. Molnár (2013: 14) considers how digital mobilization has the potential to reframe pubic space: "mobile communication technologies have in fact become powerful urban design tools that can effectively share our experience of the urban space and enlarge the urban drama while fostering new forms of sociability and reinventing public space." A further point is that these tools also provide a counter to the technologies of surveillance that are intended to prevent disruptions to the normal flow of city life. However, David Lyon (Bauman & Lyon 2013: 12) cautions that the power differential remains in that surveillance practices based on information processing track the details of our daily lives making our actions transparent, but the activities of those who conduct these surveillances become less easy to discern.

Citizens under the panoptic gaze

The panoptic gaze in fiction is not a recent development. Orwell's (1949) *Nineteen Eighty-Four* is the most frequently invoked surveillance fiction and despite its publication more than six decades ago, it anticipates several aspects that are familiar to Western societies post–9/11 – "permanent surveillance, unlimited and unwarranted detention for potential crimes, and torture" (Banita 2012: 252). These features are central to *The Hunger Games, Little Brother*, and *Omega Place* to varying degrees.

As its title implies, *The Hunger Games* plays with two key ideas – survival and entertainment. The two ideas are played out when twenty-four young people or "tributes" are compelled to fight to the death in a vast outdoor arena. The tributes are drawn from a lottery pool. When Primrose Everdeen's name is called, her sister Katniss, the protagonist in the story, volunteers to be her replacement. The other tribute from her district is Peeta Mellark. The spectacle of killing is mandatory televised viewing for the people of Panem: "the country that rose up from the ashes of a place that was once called North America" (Collins 2010: 21). The country of Panem exists as a closed world with no apparent outside communications, despite the state-of-the-art technology in the Capitol, the governing city. The technologically sophisticated Capitol with its video surveillance, smart technologies, and tracking devices is a marked contrast to the more impoverished technological landscape of many districts.

The Hunger Games extends both Orwell's and Foucault's accounts of constant visual scrutiny by fusing it with the contemporary adulation of celebrity. The outdoor arena of the Games is under constant video surveillance and manipulation by the Gamemakers. This visual panopticon is similar to Jeremy Bentham's panopticon that Foucault drew on to theorize surveillance in disciplinary societies. However, the film version of *The Hunger Games* shows more vividly than the print text how the solid architecture of the panopticon is replaced by the fluidity of electronic technologies that monitor and regulate the contestants' movements and opportunities as the Gamemakers operate a sophisticated interactive geomap of the Games arena.

The Hunger Games draws readers into its lurid spectacle of killings, rivalry between districts and individuals, and televising of intimate moments of human emotion between three tributes, Katniss, Peeta, and Rue. These viewing moments carry familiar resonances in our world of reality TV and the alternating highs and lows are judged by the viewing audience in the text who vote on whether they will send silver parachutes of food, weapons, or medicine to assist their district's contestants. "Panem" is also a reference to the "bread and circuses" (*panem et circenses*) of Imperial Rome where entertainment was used to pacify discontent or divert attention from a grievance. The text resembles contemporary Western society's appetite for voyeuristic entertainment that may be degrading and embarrassing for the participants, but offers a perverse pleasure for viewers as exemplified in the popular reality TV series *Big Brother*.

The Capitol's power to implement a surveillant vision through its fluid panoptic gaze fails to strip Katniss of her agency and capacity to take up a moral and empathic position. Surveillance technologies, along with the power brokers of the Capitol who use them, perform as the insidious antagonists in this story and one could easily argue that such a schematic alignment of evil to technological advance is a form of nostalgia for an imagined time of uncomplicated humanity and technological simplicity. Katniss realizes the importance of distinguishing herself as an individual among the mass of tributes: she plays along with pretence that she and Peeta are star-crossed lovers, thereby appealing to the sympathy of the audience. However, the Games prove to be both an ethical and moral challenge for her. She is acutely aware of the limits of the system of government in Panem and her ministering to the needs of injured tributes (Peeta and Rue) is a moral action that is at odds with the inhumane rules of the Games. Nevertheless, she does not adopt a position of moral absolutism as she finds that she must kill other tributes or be killed.

The dystopic future that is presented in *The Hunger Games* with its state-of–the-art technopolis is markedly different from *Little Brother*, which is set in a time contemporaneous to many readers, where there is more distributed access to mobile technologies, smart city transportation system, and a postoptic data-mining system of surveillance. Unlike *The Hunger Games* where control is held by the city center and the fate of the tributes is in the hands of the Gamemakers, *Little Brother* is typical of the way in which data processing is now decentralized, representing the shift from the computer to the user. User control is a key feature of *Little Brother,* which demonstrates an inherent paradox: while savvy users gain control over communication and information systems, many remain clueless as to their loss of privacy in that their personal information can be readily known and used by unknown sources. *Little Brother* reflects the new ambient intelligence (AmI) environment that recent advances in microelectronics and wireless communications are making possible. As Gadzheva (2008: 60–61) explains, an AmI environment

> implies a seamless environment of smart networked devices that is aware of the human presence and together with the ever-enhancing data mining capabilities gives the possibility for personal data to be invisibly captured, analyzed, and exchanged among countless sensors, processors, databases, and devices to provide personalized and contextualized information services.

Little Brother brings to the fore the broader implications of information control, surveillance, and privacy that come with an AmI environment. Paradoxically, it recuperates surveillance through a diffusion or democratization of surveillance systems in an attempt to redistribute power from a central control to subversive counterforces. This power at the extremities illustrates Foucault's conceptualization of power as a productive force.

The story begins when high school senior Marcus Yallow decides to cut school to go downtown to play the alternate reality game Harajuku Fun Madness. Marcus's handle is w1n5t0n, pronounced as "Winston," an obvious homage to Winston Smith, the autonomous moral agent in *Nineteen Eighty-Four.* (He changes it to M1k3y when he organizes a covert Internet resistance force against the Department of Homeland Security.) To leave school undetected, Marcus has to negotiate the school's surveillance system – the gait-recognition cameras have replaced the face-recognition cameras, which were ruled unconstitutional. As Marcus explains: "Gait-recognition software takes pictures of your motion, tries to isolate you in the pics as a silhouette, and then tries to match the silhouette to a database to see if it knows who you are. It's a biometric identifier, like fingerprints or retina-scans" (Doctorow 2008: 10).

In successfully circumventing the school's surveillance mechanisms, Marcus represents Doctorow's ideal jail breaker of electronic systems, a skill that the author believes is important for young people in order for them to retain their democratic rights (cited in Goldberg 2011: 27). However, the day takes an unexpected turn when terrorists blow up the Bay Bridge causing major death and destruction, and turning San Francisco into chaos – an event that resonates with the attack on the World Trade Center and the subsequent chaos in New York City that resulted, and the ensuing hypersecurity measures.

In taking its cue from Orwell's *Nineteen Eighty-Four, Little Brother* illustrates how surveillance technologies can affect individuals' privacy and freedom. It also shows how human beings can succumb to external forces and become passive victims or, alternatively, in Foucauldian terms, how they can become active bodies who are resistant and struggle against the limits and oppressions imposed on them, and believe in the possibility of reversal (Foucault 1976: 95–96). This capacity to resist, fight back, and overcome is the driving motivation behind Marcus's actions after he is detained in the aftermath of the bombing. Marcus finds that his carefree life as a teenager has been abruptly replaced by an existence where he is under constant surveillance. Trust in the government has been eroded and the consequent "crisis of agency" (Bauman & Lyon 2013: 146) means that normal social relations are severely constrained. For Bauman, this crisis occurs when there has been an erosion of trust in that the system of government is not working and other ways for being proactive and political need to be found.

David Lyon sees Bauman's optimism for a proactive citizenship that can be an outcome of the crisis as replacing the "hermeneutic of suspicion," that he believes characterizes our present world, with a more hopeful "hermeneutic of retrieval" which "reaches back in order to confront and engage with the present" (Bauman & Lyon 2013: 146). A similar hopefulness is characteristic of the self-reflexive quality of much critical dystopian fiction which Tom Moylan (2000: 271) suggests is potentially subversive and capable of "changing the minds of their readers": a goal that is explicitly expressed in *Little Brother.*

One side of the argument for increasing technological developments is the benefits for economic growth, security, and individual and social safety. *Little Brother* argues the other side, giving voice to growing concerns such as profiling, surveillance, tracking, and identity theft, urging readers to take action. Integral to taking action is knowing how to circumvent, deactivate, and protect user identity. The information supplied by the first-person narrator (Marcus/M1k3y) on hacking, using illegal web servers, spamming, cryptography, and radio-frequency identification (RFID) cloning is rationalized in terms of an individual's right to privacy, his/her right to know, and other constitutional rights such as the "Life, liberty and the pursuit of happiness. The right of people to throw off their oppressors" (Doctorow 2008: 201). The repeated invocation of the Bill of Rights throughout the text can be read as a totalizing conception of ideal human existence and an evocation of a pre-technological society, and the account of technological sabotage by both the state and Marcus is intended to awaken readers from any self-deceptive dreams of an ideal existence in a democratic society (an intention that is made explicit in the Afterword).

What *Little Brother* attempts is to break the ties between surveillance and a central authority and in so doing redraw the moral demarcations that organize a post-panoptical society. The ethical purchase of this text resides in its emplotment of the connectivity facilitated by surveillance and countersurveillance circuits. In response to the terrorist attack, the Department of Homeland Security (DHS) becomes a force to combat terrorism and "protect" the people of San Francisco. However, in waging its own war on terror with increased homeland security, the DHS takes on the tactics of the terrorists in the protection and security of the citizens. During his detention by the DHS after the attack on the Bay Bridge, Marcus is interrogated as a suspected terrorist. His history as an Internet-savvy user, with an antiauthoritarian attitude, is used as grounds for suspicion:

> [Marcus]: "You think I'm a terrorist? I'm seventeen years old!"
> [DHS officer]: "Just the right age – Al Qaeda loves recruiting impressionistic, idealistic kids. We googled you, you know. You've posted a lot of very ugly stuff on the public Internet." (Doctorow 2008: 41)

The interrogation of Marcus highlights the fact that after 9/11, dangerous bodies are not only those that constitute a threat from the outside, but also those on the inside who are seen as vehicles for enemies of the state. The DHS combines extreme disciplinary measures (extraordinary rendition, waterboarding, unwarranted detention) with an aggressive hypersecuritization to ensure that the surveilled body – the perceived dangerous body – is watched, tracked, profiled, and contained. When Marcus is released from detention, the DHS officer issues a warning: "We'll be watching you everywhere you go and everything you do" (p. 46). The surveillance systems are extensive – profiling based on embedded RFIDs contained in commuters' transport cards (FasTrak Pass) and other consumer goods and services, wiretapping of phone and Internet, gait monitoring surveillance, and spycams – all of which are part of the urban environment of many cities around the world. The DHS uses profiling to identify individuals as part of a group or category of persons (potential threats to homeland security) with the consequences that they lose autonomy and self-determination by being arrested, imprisoned, and tortured.

A surveilled body can also be a resistant body. This double condition occurs in the way that Marcus and his band of young hackers, the "Xnetters," manifest their Otherness by resisting and transgressing the DHS surveillance and oppressive controls. They are similar to smart mobs in their use of mobile technologies for collective action. The Xnetters use illegal, covert communication technologies to circumvent the city's security and transport systems. For example, Marcus uses the ParanoidXbox operating system to encrypt documents and communications. He also jams the tracking systems for commuters by using his arphid cloner to swap FasTrak tags with random numbers taken from cars he had passed by. In return, the DHS uses the media to spread disinformation about the Xnet movement that Marcus forms after his release. By demonstrating this capacity to act, Marcus and the Xnetters ensure that they are agential, resistant subjects that actively challenge the hegemonic domination of the DHS. Their actions are a direct counter to the loss of trust in the government (and adults generally) as well as an attempt to make it possible for their smart mob otherness to reassert itself in a form of resistance and transgression. From a Foucauldian perspective, Marcus represents a non-normative idea of freedom: a freedom that can only come with transgression.

Omega Place offers a different perspective by seeing transgression as having its own limits on personal freedom. The story takes place in UK cities but predominantly in London, a city where, ten years ago Price (2003: 37) reported, "over 800 high-resolution cameras scan about 50 arterial entryways into London, and roads throughout the congestion zone." The specificity of place is important for this text as it is given significance in terms of the human need to belong somewhere. However, place is also shown to be more than simply a physical location. Rather, it accords with Walter's (2011: 201) description that place is "a complex socio-spatial construct that can embody different meanings for different people." The title of the book appears to specify a place, but it becomes clear that "Omega Place" is a nonplace, ostensibly a code name for an eclectic group bent on destroying as many CCTV cameras in the inner city as possible, but it also has a more ominous high-security meaning. The group's Manifesto is a call for action from complacent citizens asking "What has happened to the politics of the street?" (p. 21). The Manifesto presents a view on the dystopian side of city technologies:

> Supposed to make us feel SAFER, they say. Supposed to CUT CRIME and CUT ROAD ACCIDENTS, they also say. Except we know it's really about **MONEY.** And **CONTROL.**
>
> ...
>
> **Now there's ONE CAMERA for every 14 people in the UK!** And you are being WATCHED 24/7, almost everywhere you go and whatever you are doing. (Marks 2007: 20, emphasis in original)

When seventeen-year-old, Paul Hendry leaves home after another argument with his stepfather, he witnesses a CCTV camera being smashed and after the perpetrators leave, he reads the pamphlet (the Manifesto) they have left in its place. He wonders about the logo or name on the pamphlet: "ΩP. Omega Place. Neat logo, cool name, but what does it mean?" (p. 22). Paul eventually becomes entangled with this group, and enjoys the lawless freedom that they experience – stealing cars, smashing the cameras, living in squats, smoking dope. However, it also provides him with a reflective space to think about the life he left behind. While he did not like his stepfather, and his father seems detached, Paul nevertheless realizes that life had been very different for some of the others in the group: "He thought about why he'd left home. His nice, comfortable house with a mam who would do anything for him...and a stepfather he didn't get on with. Didn't sound like much of a bad place to be" (p. 64).

There is another story that gives credence to the Manifesto's warnings and underscores Paul's naïveté. Unbeknownst to the group, the Omega Place is of interest to MI5 and a special task force has been commissioned to "close these people down, whoever they are, for good" (p. 57). Suspicions were aroused that this was not just a group of "small-time activists" (p. 56) because remotely piloted aircraft (RPA) is mentioned in its Manifesto. In contemporary society, CCTV cameras are openly visible surveillance devices, but RPAs are part of what is termed "new surveillance"[1] technologies that have expanded from military to civil applications, and have raised legal and public concerns about privacy and other civil liberties (Finn & Wright 2012). *Omega Place* alludes to these concerns through the covert operations of the special task force, which culminate in a raid on the squat, killing two members of the group.

When Paul learns of the deaths and the arrests of two other members, he asks himself, "Who the hell had allowed that to happen?" (p. 242), voicing a concern that echoes Bauman's crisis of agency. He also cannot believe that the newspapers do not report on the incident and asks incredulously, "How can two people being shot dead *not* be a story?" (p. 243). Paul considers the transience of belonging and place when he reflects upon his recent activities:

> He'd left home not knowing what he was looking for and by accident he'd found a weird kind of other family, which had taken him in and accepted him for what he was, taught him to think, taught him that you needed action *and* words.... [A]nd now, now he was lost again, the family had been destroyed.... (p. 242)

Paul's sense of being lost refers to his emotional state of not knowing where he belongs or fits in. Place, too, proves to be transient or impermanent concept to him. The squats are destroyed after the shootings, erasing any evidence of their existence. Home is where he decides to return, at least for the present.

Being lost and finding a home or a place to belong in a detached urban space are the themes of *The Lost Thing*. The story tells of a boy (Shaun) who finds and befriends a large, red "thing" that is an assemblage of mechanical and organic parts. Shaun assumes that the figure he notices on a beach must be lost and so tries to find its owner: "it had a really weird look about it – a sad, lost sort of look. Nobody else seemed to notice it was there" (n.p.). While Shaun attributes an invisibility to the thing (despite its size, bulk, and color), the reason for this "not seeing" is open to interpretation – was it fear of the unknown or disinterest that other people felt? Despite its depiction of a city and its populace appearing directionless and lacking in energy, this text offers a utopian vision in its conclusion, one "which reaches beyond a fear of the unknown to embrace new ways of being" (Bradford et al. 2008: 3).

Fear of the unknown and new ways of being are also part of the argument for and against city surveillance technologies. The city represented in *The Lost Thing* appears to have lost or long forgotten its smart credentials, if it ever had any to begin with. Information technologies are either absent or obsolete in the text, yet this is a place where a surplus of information has resulted in the formation of The Federal Department of Odds & Ends which has a pigeonhole for the abundance of useless surplus of information. Its motto is *Sweepus underum carpetae* suggesting that information is indeed redundant but does ultimately have a place – under the carpet. This motto is also a veiled reference to how information can be easily swept aside or treated as if it doesn't exist – something that Paul found out in *Omega Place* when there was no reporting on the deaths of his friends.

Information is one of two significant tropes in *The Lost Thing*; the other is entropy. The two come together in the advertisement that Shaun reads when he is trying to find the owner for the thing:

> Are YOU finding that order of day-to-day life is unexpectedly disrupted by
> Unclaimed property?
> Troublesome artifacts of unknown origin?
> Filing cabinet leftovers?
> Objects without names? (page reference)

This advertisement with its suggestion of disorder being the burden of information surplus that the average person endures daily is a reference to high entropy. As Greene (2004: 154) explains, entropy is the "measure of the amount of disorder in a physical system." High entropy equates with high disorder, whereas low entropy means the system is highly ordered:

> Entropy is defined as energy that can no longer be put to work, no longer can be organized to do something, having become chaotic, like microparticles moving out of order, aimlessly. As such, entropy is the measure of turbulence or disorder in a closed system. (Clough 2004: 9)

The end pages of *The Lost Thing* duplicate an image of seeming order – serried rows of bottle tops each separated with mathematical precision and each inscribed with mathematical formulas, scientific and technological images, words, phrases, numerals, or symbols: signs that appear random. However, one bottle top, placed in the center of the rows, is inscribed with the word "ENTROPY." We can read entropy in the illustrated end pages as a random variable among others, that contains information and that attempts an order out of the chaos of jumbled signs, symbols, formulas, and so forth. Shaun is the boy we first met on the cover of the book standing beside the empty highway with the red figure. He is a bottle top collector who randomly collects bottle tops lying on the ground but who appears to enjoy imposing order on his collection (he carries a reference book titled *What Bottle Top Is That?*, a humorous citation to the standard scientific classifactory text such as *What Bird Is That?*).

The illustrations also encode entropy. Many cite modern city images by artists John Brack, Jeffrey Smart, Edward Hopper, and others. They capture the complexity of the city space with its juxtaposition of crowds and isolated individuals, modernist architectural monuments to prosperity and empty occupancy, bored conformity and joyful chaos. The images allude to known cities. The cover adapts Jeffrey Smart's (1962) *Cahill Expressway* (Sydney) replacing the alienated modern man in Smart's work with the eponymous bulky, red hybrid "thing" and a strange-looking, slumped-posture boy (Shaun). John Brack's (1955) *Collins St., 5 p.m.* (Melbourne) appears with its orderly lines of citizens dressed in drab, sartorial conformity moving expressionlessly along the city street at the end of the workday. This image is contrasted by the almost-empty street scene of the thing and the boy who are situated outside a row of empty shops with a barber's pole in the foreground: an image that references Edward Hopper's (1930) *An Early Sunday Morning* (New York). Whereas the relaxed emptiness of Hopper's painting shows the warm bright colors of a sunny Sunday morning, when shopkeepers, customers, and commuters are still sleeping, the inclusion of Tan's two figures connotes a different kind of empty street space, where nonactivity is not suggestive of the end of the working week, but a sign of economic downturn. The general appearance of the nonplace city that Tan depicts with its rust, decay, emptiness, and emotionless citizens captures a city that is in decline, not a smart city of the future. This is an exhausted city whose buildings, monuments, and remaining infrastructure suggest it once held high hopes for a future of continued progress and prosperity.

The utopian impulse emerges when Shaun is guided to a new place by a cyborg cleaner that gives Shaun a business card with a squiggly arrow drawn on it. Shaun follows this entropic arrow,[2] a journey complicated by the profusion of similar-looking signs, other arrows, and squiggly shapes. But out of this chaos of signs he locates a buzzer which when pressed opens up a big door onto a large space of multiple colonnades, and for the first time we see a blue sky. The space appears chaotic with a proliferation of objects of different shapes and sizes, floating, anchored, assemblages of organic and nonorganic parts (Mallan 2011: 165). However, the high entropy of this space named Ut¤qIA appears to be space for many different things to coexist and contrasts with the low entropy of the government department and life on the city streets. It is also where the lost thing finds a place to belong.

In presenting the other side of the smart city story, this chapter has shown how literature can offer an insightful commentary on and explication of issues that affect the ordinary citizen with the accelerated use of electronic systems of surveillance. Literature for young people invariably is instructive and the examples discussed in this chapter offer new lessons for its readers about the changing relationship between technology, democratic freedom, and community, which can arguably be seen as foundational to the concept of smart cities. However, the social reality of cities falls short of the ideal posing ongoing challenges for citizens as well as for urban managers, designers, and policy makers. While there is a lot riding on the smart city to solve the many problems that we experience in our modern world, there is also rising concern with how individual rights and freedoms are being compromised and control over personal information is being lost through the ubiquitous presence of technological surveillance in urban environments. The combination of technology, infrastructure, architecture, everyday objects, and bodies that are part of the smart city's strategy for solving problems is also creating problems for citizens through surveillance, tracking and recording of their movements, transactions, and communications.

The adoption of advanced surveillance technologies, at both individual and societal levels, has given rise to a dramatically different kind of fixed-place panopticon that Bentham first designed. The idea of a fixed physical and geographical place in which surveillance occurs is becoming obsolete as an increasingly global and mobile post-panoptic network reconceptualizes place in terms of its flows through social and spatial arteries. The electronic panopticon is more far reaching than its predecessor's model especially when it is located in the ambient intelligence era, with the wide dissemination of RFIDs, mobile technologies, "smart" objects and surveillance tools. *Little Brother* and *Omega Place* illustrate this kind of arterial connectivity that is facilitated by surveillance and countersurveillance circuits, and highlights the diffusion or democratization of observation. *The Hunger Games* offers the insider's perspective to the moral complexity of celebrity culture and reality TV–style surveillance entertainment. *The Lost Thing* returns to the individual and the alienating prospect of a city that has lost its energy and capacity to connect. Its closing, hopeful scenario can be read as showing the value of utopian dreaming for imagining a renewed social space of belonging.

My contention in this chapter has been that fiction constructs worlds that reflect back to readers versions of the real worlds they inhabit. Texts such as those discussed in this chapter invite readers to reflect on their own world: its utopian and dystopian possibilities, and much more in-between these two extremes. The smart city is a product of utopian dreaming which is especially important as we continue to create innovative forms of security that will achieve social inclusion, economic prosperity, and sustainability, as well as come to grips with the immense possibilities made available by emergent information and mobile technologies. The dystopian nightmare of the smart city awakens us to the fears and anxieties that these same issues engender with respect to infringements to individuals' privacy and democratic rights, and feelings of marginalization and isolation. For young people fiction may provide a platform for discussing these contrasting possibilities.

References

Aarts, E., and de Ruyter, B. (2009). "New research perspectives on ambient intelligence." *Journal of Ambient Intelligence and Smart Environments*, 1, 5–14.

Banita, G. (2012). *Plotting justice: Narrative ethics and literary culture after 9/11.* Lincoln, NE: University of Nebraska Press.

Bauman, Z., and Lyon, D. (2013). *Liquid surveillance: A conversation.* Cambridge, UK: Polity Press.

Bradford, C., Mallan, K., Stephens, J., and McCallum, R. (2008). *New world orders in contemporary children's literature: Utopian transformations.* Houndmills, UK: Palgrave.

Clough, P. T. (2004). "Future matters: Technoscience, global politics, and cultural criticism." *Social Text*, 22(3 80), 1–23.

Collins, S. (2010). *The Hunger Games.* London: Scholastic Children's Books.

Debord, G. (1988). "Comments on the society of the spectacle." Retrieved November 14, 2013, from http://libcom.org/files/Comments%20on%20the%20Society%20of%20the%20Spectacle.pdf

Doctorow, C. (2008). *Little brother.* London: Harper Voyager.

Finn, R. L., and Wright, D. (2012). "Unmanned aircraft systems: Surveillance, ethics and privacy in civil applications." *Computer Law & Security Review*, 28, pp. 184–194.

Foucault, M. (1975). *Discipline and punish: The birth of the prison* (A. M. Sheridan Smith, trans.). Harmondsworth, UK: Penguin.

Foucault, M. (1976). *The History of sexuality: Volume 1* (R. Hurley, trans.). Harmondsworth, UK: Penguin.

Foucault, M. (1980). *Power/knowledge: Selected interviews and other writings* (1972–77) (C. Gordon, ed.). New York: Pantheon.

Gadzheva, M. (2008). "Privacy in the age of transparency: The new vulnerability of the individual." *Social Science Computer Review*, 26(1), pp. 60–74.

Goldberg, B. (2011). "Privacy worries grow more public." *American Libraries*, 42(5/6), pp. 26–27.

Greene, B. (2004). *The fabric of the cosmos.* London: Penguin.

Mallan, K. (2011). "All that matters: Technoscience, critical theory, and children's fiction." In K. Mallan and C. Bradford (eds.), *Contemporary children's literature and film: Engaging with theory* (pp. 147–167). Houndmills, UK: Palgrave.

Marks, G. (2007). *Omega place*. London: Bloomsbury.
Molnár, V. (2013). "Reframing public space through digital mobilization: Flash mobs and contemporary urban youth culture." *Space and Culture*, XX(X), pp. 1–16.
Moylan, T. (2000). *Scraps of the untainted sky*. Boulder, CO: Westview.
Orwell, G. (1949). *Nineteen eighty-four*. London: Secker & Warburg.
Price, A. (2003). "Surveillance and the city." *The Next American City*, 2, pp. 36–38.
Rosen, J. (2003). "A cautionary tale for a new age of surveillance." In P. Griset and S. Mahan (eds.), *Terrorism in perspective* (pp. 303–312). London: Sage.
Summers, J., and Winefield, H. (2009). "Anxiety about war and terrorism in Australian high-school children." *Journal of Children and Media*, 3(2), pp. 165–184.
Tan, S. (2000). *The Lost thing*. Melbourne: Lothian.
The Hunger Games. (2012). Directed by Gary Ross. USA: Lionsgate.

Notes

1 RPD are also known as UVA (unmanned aerial vehicles) and RPV (remotely piloted vehicles). The UVA was so named in Surveillance Studies Network's submission to the UK House of Lords. (See House of Lords Select Committee on the Constitution. Surveillance: Citizens and the state, vol. 2. HL paper 18, second report, session 2008–09. London: House of Lords; 6 Feb 09; cited in Finn & Wright 2012: 185.)

2 The second law of thermodynamics offers "an arrow in time." A forward in time arrow points in the direction of increasing entropy (see Mallan 2011: 162–163).

10

Smart City Needs Smart People: Songdo and Smart + Connected Learning

Michelle Selinger and Tony Kim

Abstract

Newly industrialized countries in the Far East are characterized by the rapid growth of new cities, and Songdo in South Korea is one. Cisco's Internet Business Solutions Group for whom the two authors previously worked, were integrally involved in helping to develop a strategic plan for the city to become a sustainable model of a "smart and connected community" and to become a hub for the Asia-Pacific headquarters and the international business of multinational firms creating clusters interconnecting industry, academics, and R & D. Cisco is engaged in the implementation of the city's pervasive network infrastructure, which will support the city's aspirations to rethink the way cities are designed, built, managed, and renewed to achieve economic, social, and environmental sustainability using the network as the fourth utility.

Songdo planners recognize, and are planning, an education system that will produce active and engaged citizens with the skills and disposition to work, live, and learn in a connected environment. Additionally, Songdo residents have access to information and services that enrich their lives, with solutions for their home, education, and transportation through video-based service channels. The implementation of networked collaboration tools and video technologies in education, as in other areas of the city's activity, will help ensure that the new city's economy will develop and grow. The Songdo story exemplifies what can be made possible through the effective use of technology across various sectors and community development of the city.

IFEZ and Songdo

Developed on reclaimed land from the Yellow Sea, Incheon Free Economy Zone (IFEZ) is an outpost for international business to support and guarantee optimal economic activity. IFEZ is the manifestation of Korea's vision for international recognition and designated by the government as Korea's first free economic zone in October 2003. Its total projected population is 512,000 in a total site area of 132.9 km^2 with an infrastructure development budget of 21.45 trillion KRW (US$20 billion) (IFEZ, n.d.).

IFEZ consists of Songdo, Yeongjong, and Cheongna International Cities including Incheon International Airport/Ports as the hub for government's strategy of making Northeast Asia's economic central. It is located at the center of the Northeast Asia economic zone. Northeast Asia makes up one fifth of the global economy, including major economic countries such as China, Japan, Hong Kong, and Russia. In Northeast Asia, over 1 million people reside in 61 foreign cities that can be reached within a three-hour flight, enabling rapid adaptability to Asia's dynamic business market. Songdo is designed as a city located in south Incheon, equipped with a competitive advantage for international businesses, information technology, and biotechnology (IFEZ, n.d.).

Vision and aspirations of IFEZ

In an attempt to increase foreign investments and local hiring based on high-tech and service industries, Incheon Metropolitan City developed IFEZ. This financial strategy was grounded in the accessibility of Incheon International Airport, which serves 87 international airports and 176 cities, and Incheon's port. Incheon Metropolitan City focused its structural reform on a vulnerable industrial ecosystem of small-scale manufacturing, and constructed IFEZ with a vision toward a sustainable model and shared growth between business and society.

A "compact" city is an ideal city of the future in which business and life thrive side by side and where all aspects of everyday life are made easily accessible. Business and high-tech industries, health care, education, culture, tourism, leisure, and shopping are conveniently located in one compact, well-designed area.

Songdo Smart + Connected Community

Many cities do not have interoperable systems and protocols across an urban center are not interoperable. Converging these systems onto integrated open systems–based network-enabled infrastructure creates significant opportunities for productivity, growth, and innovation. Cities of the future and many innovative cities are addressing the urban challenges and opportunities by thinking about the network as the fourth utility. So on this basis, in 2009 Songdo launched a "smart + connected community" initiative (Cisco, 2011a).

Cisco Smart + Connected Communities use intelligent networking capabilities to weave together people, services, community assets, and information into a single pervasive solution. "Smart + connected" acknowledges the essential role of the network as the platform to help transform physical communities to connected communities. It also encapsulates a new way of thinking about how communities are designed, built, managed, and renewed to achieve social, economic, and environmental sustainability. Cisco's innovations in Songdo include the signature TelePresence system, an advanced videoconferencing technology that allows residents to access a wide range of services including remote health care, beauty consulting, and remote learning, as well as touch screens that enable residents to control their unit's energy use.

The Songdo Smart + Connected community blueprint is anchored on a broadband infrastructure which enables new smart services for work, learning, health, travel, video-enabled public services, and smart green buildings and homes.

People in Songdo have access to information and services that enrich their lives, with solutions for their home, schools, and transportation through various network-based service channels and user devices such as smartphones, TV, in-home display, and computers.

The education landscape in Songdo

One of the most attractive aspects of Songdo is its human resources. Songdo is optimally located in a metropolitan area and is fast becoming a hub for education by establishing international schools, Songdo Global University campuses, Yonsei University's international complex, and a domestic university campus for business and research.

Songdo Chadwick International School for K–12 opened in March 2010 with 2,080 students, and major facilities such as a digital library, assembly hall, concert theater, sports center, all of which are fully connected with video-enabled systems and broadband convergence infrastructure. The school is the first of its kind in Korea to offer a world-class international school program that local Korean students can attend together with expatriates and dual-citizenship holders.

The school offers a smart and connected educational environment where foreign investors can safely educate their children through programs that apply world-class curriculums operated by Chadwick International School. An additional international school will be built in Songdo in the second phase of IFEZ development project.

Incheon is currently building the Songdo Global University Campus where international educational institutions of excellence (four-year universities, graduate schools, affiliated research institutes) have been recruited to this extended campus zone, where they will provide students with opportunities to acquire the same academic degrees that are issued at the participating institutions in their home countries. The Songdo Global University Campus is designed as Korea's first initiative to develop a joint campus with foreign educational institutions. The aim is to develop the campus into a cluster of industry–academia–research institute collaboration as well as a Northeast Asian hub of education.

Songdo Global University Campus is a developing education and research hub complex that is responding to increasing demands for international education, construction of knowledge-based industry and infrastructure, and cultivation of global professionals. On the Songdo Global University Campus, State University of New York (SUNY) Korea opened in March 2012, followed by George Mason University, the University of Ghent, and University of Utah. SUNY Stony Brook, North Carolina State University, University of Southern California, and University of Delaware are about to take up residence to offer a comprehensive and first-class university experience through the integration of competitive degree programs supported by a next-generation pervasive video-enabled learning environment. These video facilities will support links with the home institutions and provide opportunities for shared teaching and peer engagement across satellite campuses in countries like China, India, Malaysia, the Middle East, and the home campus. Not only will this support greater international collaboration, but it will also ensure that wherever students are located, they will have access to the best and most relevant teachers from their institution for the programs they are studying, and specialised support for their theses and dissertations.

Innovations in communications and collaboration technologies have made a significant impact on the way university partnerships and overseas campuses have taken shape around the world. For example, when Qatar Education City (Qatar Foundation, n.d.) was being established in 1995, the Qatar Foundation offered huge incentives to the universities they were trying to attract. This included the incentive to relocate some of their best professors to teach locally. Now with ubiquitous video technologies that are available and with good broadband connectivity, this imperative is no longer essential for the success of satellite campuses and the equity of student experience. Higher education institutions are able to share faculty across campuses that are located almost anywhere in the world.

But it is not just U.S. universities that are locating satellite campuses in Songdo: Yonsei University, one of the leading universities in South Korea, is developing its global academic complex in Songdo, with a total site area of nearly 100,000 m^2 opened in March 2012. Yonsei University's R & D complex cluster is providing flexible links to foreign universities and companies. It is also planning to fully utilize video-based advanced learning features on a smart campus that consists of high-tech industrial cluster and anchor facilities.

Smart + Connected learning in Songdo

Songdo has built a ubiquitous ICT infrastructure throughout the city that will be the platform for the development of a highly connected learning environment. Songdo's advanced video-based communication technology on a ubiquitous sensing and networking infrastructure to manage schools, universities, public facilities and spaces, and so forth, will lead to a new and unprecedented level of convenience and opportunity for learning. Through a video-enabled ubiquitous learning environment in Songdo, every learner has easy access to smart + connected learning services anywhere and anytime. In the smart + connected learning environment of Songdo, high-quality video links facilitate real-time distance education and training, collaborative meetings and participation of guest speakers from geographically dispersed areas at lectures, seminars, workshops, and conferences.

A good example of how smart + connected learning is developing in Songdo can be found at the Chadwick Songdo International School described above, which has been built in the Songdo International Business District. The school has an exceptional 500,000 ft^2 of state-of-the-art facilities and ubiquitous infrastructure to support its educational programs (Heren, 2011). Class size is around 20 students to a class, compared to 25–30 in standard Korean schools. However, not all the classes are small—many classes are conducted for large groups of students who all need to learn the same ideas.

The Chadwick School was first opened in California more than 75 years ago by Margaret Lee Chadwick. The Chadwick School model has "an educational philosophy based on liberalism and creativity while having academic excellence and development of exemplary character as its goal" (Heren, 2011). Indeed such is the model's success that 83 percent of its graduates were admitted to the top 10 percent of U.S. universities. Many parents doubted the success of a Chadwick School and the application of its somewhat alien pedagogical model in Korea—especially the notion of an inquiry-based curriculum and teachers who were interested in their "students' thoughts" (Heren, 2011, p. 3).

Traditional schooling in Korea is characterized by a "cramming" system of education. Recently this approach has led to many angry protests from parents wanting changes in the existing education system. After all the complaints about cramming schools and the need for independent and creative thinking, education leaders in Korea are making attempts to break away from the spoon-fed and short-sighted approach to education of the past toward a new approach in the classroom to graduate students capable of facing the new challenges of the 21st century. Study loads for each subject have been reduced to an appropriate level, while curricula that accommodate different needs of individual students have also been introduced. Independent learning activities designed to enhance the self-directed learning required in the knowledge-based society have either been introduced or expanded (MEST, 2011).

In order to fit with the smart + connected learning theme, some of the principles of the Chadwick School appear to align with the aspirations for the city. Smart cities need smart people and if students are to succeed and help position Songdo in its place in an increasingly globalized society, and if Songdo is to live up to its promise, then a change in the way students are taught might certainly be deemed necessary if the city is to realize its future as a smart + connected community.

The Songdo International School had a unique opportunity to design the facility so that every space can promote learning. There are learning opportunities, materials, displays, and technology throughout the building in passageways, cafeterias, offices, and other public spaces in addition to instructional space. The facilities are arranged in three phases of schooling to include Elementary School, Middle School, and High School, and there are shared facilities buildings. Every classroom is physically connected to the other classrooms, supporting the concept of collaboration throughout the entire campus. There are also dedicated "collaboratories" on every floor of the building to encourage learning outside the regular classrooms, and there are TelePresence rooms and video-conferencing facilities in classrooms. All the facilities were designed with security and transportation flow in mind, with separate areas for student dropoff and visitor access. The abundance of state-of-the-art technology available to students at Chadwick International offers unparalleled opportunities for personalized instruction and self-directed learning, while the focus on community provides a supportive environment for all students (Warmington, n.d.).

It took some time for the English teachers to live with laptops in the classroom, but they have found that the cross-referencing ability through high-quality video conferencing link with "Community + Connect" environment makes their discussion more interesting. The online environment at Songdo global campus encourages access to multilingual and multicultural resources and connections, and encourages students to apply their learning from these interactions to their projects.

With pervasive video across the city, the opportunity for links to the community, to business and industry, and to the universities offers students an opportunity to add authenticity and relevance to their education. The school's walls are blurred and teachers will need to take on new roles to manage the interactions students have with the world outside and monitor the impact this has on their learning. Teachers will need significant professional development to help them recognize that their roles will change from being dispensers of knowledge to become orchestrators of the learning environment.

Professional development at Songdo International School prepares teachers for their new roles in the school by perceiving teachers as learners as well as students. Teachers do not wait to be trained on the use of new technologies in their field; instead, they seek out the latest developments and learn how to use them. The school has assembled a directory of online courses and tutorials relevant to their field, to which they send students. Teachers at Songdo International School are prepared to deliver just-in-time learning—when a situation arises that demands the application of a new skill or concept, the teacher is ready to help students learn it. Teachers take on the role of teaching coach to project groups, which provide many opportunities for this kind of learning, and the careful design of group topics helps to guarantee that the most important of the concepts that need to be understood and learned rise to the surface.

In addition to providing a fully illustrated textbook online for each subject, the school provides an extensive library of electronic texts that can be downloaded to students' laptops or to their tablets, formatted for ease of reading on these ubiquitous portable devices. The library at Songdo International School is no longer just a place to store books; it has become the hub of the school, with spaces designed especially to facilitate small-group project discussions that have become an important mode of learning in the school.

The school provides high-quality videoconferencing capabilities including Cisco TelePresence (Cisco, 2011b), which ensures that distance-learning opportunities can be fully supported in numerous ways. As it is an international school, there are students who may have to leave the school for extended periods of time in order to visit their home countries for a variety of unavoidable reasons. Video and other collaboration technologies ensure that students are still connected to their classes, albeit at a distance and are able to continue their studies and keep pace with their peers. Their sense of isolation is reduced, as they are able to maintain rapport with their teachers and their classes. Additionally, subject-matter experts, guest speakers, and remote teachers make regular appearances in classrooms and at TelePresence rooms, extending the human resources available to students as they learn and also providing students with access to experts, practitioners such as local business and community individuals across the pervasive network in Songdo, but also to communities, business, scientists, museum curators, artists, authors, and a host of experts, not just in Korea but across the world.

One example in the school is for teaching English. The English teacher uses video streaming so that students can hear English dialogue and they connect via videoconferencing or TelePresence to certified English teachers in the United States and Canada for students to engage in a 15- to 20-minute discussion with certified teachers who are native English speakers. The lesson concludes with the Songdo-based teacher summarizing what has been learned and checking for understanding. This is coteaching in a connected world with one teacher present and the other virtual. It provides students with access to clear pronunciation, the opportunity to learn colloquialisms, and to test out their speaking skills in a real-time conversation with a remote native English speaker.

This access to authentic voices, which supplement the teacher's and provide variety in the classroom, are strong motivators for learning. Studies, such as that demonstrated by the Cleveland Clinic in Cleveland, Ohio, have indicated that students taught aspects of the science curriculum via live video links to operating theaters in which surgeons talk through operations they are performing, such as open-heart surgery and laparoscopic surgery, have proven strong motivators for students and demonstrated improvements in both learning and life aspirations (Strickland, 2010). In Australia, the Connected Classrooms Project has achieved similar results with young learners becoming more interested and excited about literature through access to authors in a specially prepared video-enabled authors week, or by finding out more about dinosaurs through a video conference with a museum curator accompanied by authentic artefacts that cannot leave the museum (New South Wales Department of Education and Training, 2010).

Discussion

The interoperability and connectedness of Songdo has the potential to demonstrate a new paradigm for education in which pervasive video enables communication and collaboration in unprecedented ways. But what is the research evidence that supports the effectiveness of this approach?

After closely reviewing more than 100 studies specifically related to research into all forms of video, Cisco-commissioned research (Greenberg and Zanetis, 2012) identified almost 50 that were drawn on for a meta-study. The pedagogical impact of the many faceted forms of video suggested by these studies is summarized by three key concepts: interactivity with content, engagement, and knowledge transfer and memory. These are part of a continuum in which interactivity with content becomes the key principle and a means for cognitive development: the learner interacts with visual content, whether verbally by note taking, by thinking or by applying concepts. Engagement consists of the learner's connection to visual content—the ways in which a learner becomes drawn in by video, whether on demand or real time, narrative or pedantic. That interactivity and engagement begin in the affective realm, the feeling side of learning. Once engagement occurs, the continuum then flows into memory and knowledge transfer: the learner, according to some studies, may remember better.[1] The net result in theory is a combination of affective and cognitive development, and retention of content.

The use of video for informal conversations has increased with the increase in bandwidth, video compression technologies, and Voice over IP applications such as Skype. The video component of such applications is now integral and most Internet-enabled devices come with built-in webcams and laptops, and phones and tablets with built-in cameras. A study by Pew in the United States showed that of the 73 percent of American adults who were Internet users, 23 percent have participated in video calls, chats, or teleconferences, and that among the 85 percent of American adults who have cell phones, 7 percent have used their phones for video calls, chats, or teleconferences. This equates to 19 percent of American adults who have either used the Internet or their cell phone to participate in video calls—and in many cases, people have used both technologies for video chats (Rainie and Zickuhr, 2010, p. 3). The implications of this is that video is increasingly becoming a ubiquitous communication mechanism and both students and teachers will come to expect the use of such technologies in all facets of their lives for communication and collaboration.

Wainwright (2004, p. 5) in a review of the research on the impact of videoconferencing on distance education showed

> Unequivocally, that two-way, interactive video conferencing technology can be an extremely effective medium for delivering quality education to a broad, geographically dispersed student population. The research clearly shows that the technology has helped governments address mandates for economic and infrastructure development (not to mention internal agency training), helped universities follow mandates for educational outreach, and helped colleges, universities, and secondary schools reach out to vastly expanded student populations while also finding new sources of content and expertise. It also tells us when it is practiced well and when not. Like any technology, it can be abused, misused, inappropriately applied, or fall into neglect if not deployed with proper planning and training.

This research suggests that *appropriate* use of videoconferencing can have a beneficial effect on learning, and therefore some work needs to be undertaken to define what "appropriate" means for teaching and learning in the classroom as well as for distance and virtual learning.

Streaming video

So far, the focus of the discussion has been on videoconferencing, and of course streaming video will be of importance as dedicated video channels are streamed to educational institutions, businesses, and the home for both information and learning. The more that teachers use video content, the more benefits they tend to see. Percentages of teachers finding value in multimedia and video content have increased each year since 2007 (Grunwald Associates LLC, 2010).

In the 2010 survey,

- 68 percent believe that video content stimulates discussions;
- 66 percent believe video increases student motivation;
- 62 percent believe video helps them be more effective;
- 61 percent believe video is preferred by students; and
- 47 percent believe video directly increases student achievement.

New secure portals for streaming video increase the value of the resource as they allow for collaboration on the content where it is lecture capture, a documentary, or a demonstration. Products like Cisco's Show and Share (Cisco, 2011c) provide not only a secure environment to upload and store video, but also allow for easy searching of specific content, related video, and authors. Show and Share, for example, has a strong asynchronous collaboration element through time-line markers where authors and viewers add questions, place pointers to related resources, add comments, or have a threaded debate at the relevant point in a video. Video in this portal also has the capability for speech-to-text translation so that finding a particular video or a section of the video is made easy and reduces the need for editing.

The functionality of such video portals for education makes one-way streaming video far more interactive in a number of ways:

1. Video as the focus for content and collaboration

Rather than embed video in text, the freeze frame functionality means that at strategic points

- teachers put in questions for discussion or consideration, and links to other resources including text and other multimedia; and
- students ask questions on a point made in the video that anyone can answer or it can be further discussed.

2. Video for teacher professional development

Comparison of practice between teachers of different experience, and different views of teaching can be observed and discussed.

- Threaded discussions at salient points where teachers can ask questions, point to behaviors or characteristics of excellent teachers, and ask student teachers and others on professional development to discuss in the context at the appropriate point in the video
- Videos showing the same class being taught by different teachers so those undertaking professional development can analyze different teacher behaviors, which they can try out (and video themselves) in their own classrooms in order to develop their own teaching style
- Watching clips of particular classroom techniques, for example, questioning, response waiting times, formative assessment, use of ICT embedded in teacher practice, classroom management, and so on.
- Student teachers recording and observing their own classroom practice and revising before teaching a class, and for reviewing after teaching a class

3. Halfway between a video conference and streaming media delivery

Video conferencing is synchronous collaboration. Streaming video, podcast, and vodcast is delivery, while a collaborative video portal is asynchronous collaboration.

4. Students and teachers as creators of content

Students as documentary filmmakers, which requires problem solving, critical thinking, planning, collaboration, group work, assessing the outputs, and peer review, and can be developed as a cross-curricular activity. The video is searchable and available for comment, peer review, and rating, so it can be considered as part of the assessment process.

Teachers and students record a practical experiment or demonstrate practical skills for students to copy, as well as record traditional lectures that are searchable through tagging and speech-to-text translation, so that students need only find the parts they need to revisit to ensure or check for understanding.

5. Alternative forms of assessment

Students choose how to demonstrate their understanding, thus promoting oral communication and suiting learners from different cultures, particularly those where the oral tradition is more prevalent than text (Selinger, 2004). Students can collaborate on production virtually and face to face.

Student learning is not only assessed in new ways beyond the traditional essay or multiple-choice assignment, but the process of filmmaking involves demonstrating those essential skills students need for the next stage in their life trajectory such as critical thinking, making choices, reasoning, collaboration and teamwork, problem solving, and so on. Through video filmmaking, students also demonstrate skills of sequencing, being succinct, presenting an argument, and storyboarding. And they can add links to other documents as further evidence of learning at strategic points.

Teachers, peers, and others use the portal functionality to comment on students' videos which are then revised and edited.

6. Video evidence of learning

Teachers record oral presentations, which are presented for external and peer evaluation and with comments made at marked points on the video. The inbuilt rating systems are used for peer and teacher evaluation. The best videos can then be used as future course content.

7. Preparing and rehearsing a lecture or presentation

With portal speech-to-text translation tools, teachers or students are able to check coherence, and video editing features mean a sequence can be re-recorded and added in.

Next Steps and Suggestions

Songdo, an area that has been reclaimed from the sea, is now gearing itself with the future outlook of an international city. It is expected to be the catalyst of growth for Korea's new generation as its residents prepare for tomorrow with their dreams full of hope and passion. When the Songdo smart + connected community project is completed, the city will be able to monitor carbon emission levels and intelligently manage urban resources, information, and carbon footprints. And the city will deliver video-enabled services from learning, health care, transportation, building management, energy and water management, to safety and security while reducing carbon emissions by 30 percent compared with other newly developed cities in Korea.

Every learner will fully enjoy the Songdo smart + connected learning environment through the use of a variety of tools for school work, including video production and video-enabled collaboration. Because connectivity and video facilities are available across the city, learning can be linked closely to the needs of the individual, the community, and the businesses that inhabit the region and who will be attracted to the region because of the capabilities of its citizens.

Learners will develop the skills they need to succeed in the future, and the ubiquitous connectivity enables them to extend their learning beyond the walls and the length of the day in the educational institution. The true global city will function better if there is a smart + connected learning environment attracting global talent. Songdo will mark the success of an international public–private partnership by building the true international business hub that shows an example of how "smart cities need smart people."

Note

1 Some debate exists on memory enhancement. Most studies believe visual content helps learners remember concepts and ideas and practices; a few disagree.

References

Cisco. (2011a). *Smart + Connected Communities*. http://www.cisco.com/web/strategy/smart_connected_communities.html

Cisco. (2011b). TelePresence. http://www.cisco.com/en/US/partner/products/ps7060/index.html

Cisco. (2011c). Show and Share. http://www.cisco.com/en/US/partner/prod/video/ps9339/ps6681/show_and_share.html

Greenberg, A. D., & Zanetis, J. (2012). *The impact of broadcast and streaming video in education: What the research says and how educators and policymakers can begin to prepare for the future.* San Jose, CA: Cisco Systems.
Grunwald Associates LLC. (2010). "Deepening connections." PBS Annual Teacher Survey on Media and Technology. Bethesda, MD.
HEREN. (2011). "All about the international schools." Available at http://www.chadwickinternational.org/uploaded/About_Us/Press_room/0209Chadwick_Heren_article(Eng).pdf.
IFEZ. (n.d.). *About IFEZ.* http://www.ifez.go.kr/eng/en/m1/ifez/screen.do
IFEZ. (n.d.). *About IFEZ Education.* http://www.ifez.go.kr/eng/en/m2/edu/screen.do
Incheon Metropolitan City. (n.d.). *About Incheon.* http://www.incheon.go.kr/icweb/html/web39/039001001.html
MEST. (2011). *Education for the future.* http://english.mest.go.kr/web/1692/site/contents/en/en_0203.jsp
New South Wales Department of Education and Training. (2010). *Connected classrooms program in action.* https://www.det.nsw.edu.au/detresources/ccp_in_action_compendium_FNOouLXKim.pdf
Qatar Foundation. (n.d.). *Qatar education city.* http://www.qf.edu.qa/
Rainie, L., and Zickuhr, K. (2010). *Video calling and video chat.* http://pewinternet.org/Reports/2010/Video-chat.aspx
Selinger, M. (2004). The pedagogical and cultural implications of a global e-learning programme. *Cambridge Journal of Education*, 34(2), pp. 213–229.
Strickland, R. (2010). Cleveland Clinic Foundation.
Wainwright, A. (2004). *Navigating the sea of research on video conferencing-based distance education: A platform for understanding research into the technology's effectiveness and value.* http://wainhouse.com/files/papers/wr-navseadistedu.pdf.
Warmington, R. C. (n.d.). *From the president.* http://www.chadwickinternational.org/page.cfm?p=199.

11
Experiments with Smart Zoning for Smart Cities

Hassan Arif, Roland J. Cole, and Isabel A. Cole

Abstract

A city that has adopted the technology that makes it a "smart city" can use "smart zoning" to achieve its land-use goals. We believe that to make smart cities work, we need them to be more "variably" dense than they are. However, regardless of the land-use goal, smart technology can support the use of smart zoning to achieve that goal. We define smart zoning as using technology to (1) specify outputs rather than inputs; (2) use formulas rather than specifications; and (c) request, gather, and analyze citizen input on goals and particular zoning decisions. Smart zoning has the potential to provide a more flexible model of zoning, responsive to public needs and demands, than traditional "Euclidean" methods of zoning.

Cities need an experimental approach in urban planning whatever their goals, and especially if those goals are to allow, even encourage, many experiments that mix high and low density, residential, commercial, recreational, and even industrial uses. This includes more vest-pocket parks and areas simply left "wild" for as long as possible. We identify some of the experiments throughout the world already taking place along these dimensions, with zoning laws and prizes encouraging experiments that mix public and private spaces, high and low density, and the "accidental mingling" that leads to smart collaboration, innovation, and improvements in quality of life.

We examine in more detail a few experiments with smart zoning activities in several cities in the United States and Canada to illustrate some of the challenges and opportunities of the smart zoning approach. We conclude with some suggestions for cities that want to try a smart zoning approach.

Introduction

Information and communications technologies (ICTs), ranging from the Internet to social media to programming algorithms, offer new frontiers in public administration, law, and urban planning. ICTs can create new avenues for citizen input, whether through online means of ascertaining public views via municipal websites, or through social media such as Facebook and Twitter where residents can express their views and often directly interact with municipal officials. ICTs can provide the means to develop bylaws and regulations which are more flexible: monitoring such as noise sensor technologies can allow adaptation in real time, mapping technologies can ascertain attributes such as pervious land (land that absorbs rainwater) and aid in adjusting the regulation of one site based on the attributes of neighboring properties.

This adaptability and flexibility, this facilitation of public input and access, potentially represents a move away from top-down modes of governance, away from broad-stroke restrictions that either subsume local variations and realities or require site-by-site variance procedures. It can represent a move toward a model of governance more in tune with the variable nature of reality. Smart zoning – under the rubric of smart cities – offers a means to help realize this potential.

Smart zoning entails the application of ICTs to zoning, building codes, and land use. It involves, where feasible, adjusting to changing and unique conditions in neighborhoods – outputs – rather than broad regulations limiting activity – inputs. Formulas and algorithms can provide the programing needed to monitor and adjust to outputs such as changes in noise level. A smart approach to zoning can provide more flexibility in implementation of bylaws and regulations, allowing for more diversity in building type and streetscapes through a more variable and "on the ground" method of regulation with adaptation to local circumstances and facilitation of citizen engagement (Smart Code Version 9.2, 2013). Smart zoning practices – through social media – can facilitate greater dissemination of information and transparency of municipal practices. This chapter illustrates these benefits, through a general conceptual discussion and an examination of applications of smart zoning practices – or at least practices analogous to smart zoning.

These applications are not perfect illustrations of smart zoning. We could not find any in our research. However, they illustrate how ICTs can be used by municipalities to develop and implement better policies, to base decisions more closely on demands and situations "on the ground" accounting for the unique dynamics of different neighborhoods and locations. This allows more flexibility and variation in policy for different localities. This can be seen as a contrast to "nonsmart" methods of zoning which lack this flexibility and can lead to overbroad applications which subsume local differences. We urge cities to explore the potential of ICTs to enhance policy and decision making and its limited smart zoning–like applications.

Background

Shortcomings of current "nonsmart" zoning

How is smart zoning different from current "nonsmart" zoning practices? In the United States, post–World War II zoning is largely "Euclidean." This name comes from a case heard before the U.S. Supreme Court involving the city of Euclid, Ohio. The court upheld the city's zoning of a sizable area for single-family detached homes only.

Euclidean zoning is based on two premises: first, that landowners are treated most "equally" when each in a zone is allowed/required to use its plot the same as every other plot; second, that private and social value are maximized when all lots adjoining a lot are used the same way. These two premises often lead to larger zones as such zones have relatively few lots bordering the next zone. Also, these two premises can lead to an increasing number of specific uses – that is, from single-family only to minimum lot sizes, to minimum house sizes, to "architectural approval for house style" and how many unrelated individuals can live in a given house. Also, until outlawed for reasons outside municipal jurisdiction, these premises can lead to restraints based on race and ethnicity. Defenders of Euclidean zoning will point out these results, vast areas of single use and an increasing number of specific uses, are not necessarily always the outcome of Euclidean zoning.

Critics say that Euclidean zoning produces bad results that leads to monotonous car-dependent landscapes that neglect the human element. One critic, urbanist Jane Jacobs, stated that to thrive, cities needed high densities of people and activities, mixtures of primary uses as opposed to single-use zoning, small blocks and pedestrian-friendly streets, and a mixture of older and newer buildings. Euclidean zoning leads away from these goals by promoting lower densities, single-use areas, car-dependent roads and landscapes, and separation of old and new buildings (Wickersham, 2001). Other critics, like Gardner (2014), state that Euclidean zoning frustrated planning, giving too much deference to wealthy landowners and too little to regional and environmental concerns. Burdette (2004) states that monotony and sprawl are problems from this zoning. Inniss (2007–2008) writes that Euclidean zoning treats the city as a machine rather than as a living organism.

Karkkainen (1994) criticizes current Euclidean zoning practices for favoring certain property owners at the expense of others, for being exclusionary, for adding unnecessary transaction costs, and for producing inefficient decisions in land-use allocation. Other criticisms of Euclidean zoning include its production of food deserts – spaces far from grocery stores and restaurants – and play deserts – spaces far away from parks and playgrounds. Pogodzinski (1991) counters these criticisms stating that such results reflect underlying politics rather than zoning itself. Inniss (2007–2008) writes that alternatives to Euclidean zoning, such as form zoning or performance zoning, may produce equally bad results even if of a different variety. Form-based zoning replaces most limitations on different uses for buildings with strict limitations on the size and shape of buildings (Schleicher, 2013). Performance zoning is what we have called "output zoning"; it focuses on the outputs likely to occur – noise, pollutants, traffic, light, and so forth (Ottensmann, 1999–2000). The general goal of these non-Euclidean alternatives is to allow "compatible" mixing and density higher on average, but also with more variability (some areas very high, some very low) (Schleicher, 2013).

While ICTs can be employed for multiple purposes, employing them for smart zoning methods could be used to counter the shortcomings of Euclidean zoning. It can also be used to counter criticisms of form-based zoning and performance zoning not achieving objectives of variability, walkability, and interesting/varied neighborhoods. Smart technology, especially things like satellite mapping and geo-coding, could be used to support all forms of current zoning. However, smart technology makes alternatives more feasible, not that it makes them inevitable or only makes sense for alternative zoning schemes. Smart zoning can provide a means to facilitate greater public input and on-the-ground information to make better policy decisions (Ottensmann, 1999–2000).

The focus of this chapter will be on the opportunities smart technology offers to move beyond the limitations of Euclidean zoning, to adapt zoning, starting from any form, to both a process and a system that will be more effective at achieving community goals.

Before exploring examples of smart zoning – including policies roughly like smart zoning – we consider concepts related to smart zoning. These are smart regulation and wikigovernment. Both concepts are based on similar principles of adaptability and flexibility and describing them provides a context for the discussion of smart zoning. First, though, an outline of the basic principles behind smart zoning.

Smart zoning – basic principles

Smart zoning involves using ICT (smart technology) to move beyond fixed input rules toward something more flexible and responsive, reflecting the dynamic and changing characteristics of cities and neighborhoods. The basic principles of smart zoning are output zoning, formula zoning, balance zoning, and stealth zoning.

The first, output zoning (sometimes called "performance zoning" or "impacts zoning"), refers to the specification of outputs – such as noise levels in real time – rather than inputs (the activities involved) (Types of Zoning Codes, 2014). The latter is the traditional method of zoning law. The second principle, formula zoning, refers to the situation where the authority specifies formulas, such as "coverage impermeable to rainwater absorption cannot exceed X percent" rather than fixed specifications such as "side setbacks must be equal to or greater than ten feet." The third principle, balance zoning, involves balancing different sites against each other, and is key to adjusting to unique local conditions rather than employing a city-wide blanket regulation. Related procedures are sometimes called "forms-based zoning," "floating zones," or "cluster zoning" (Types of Zoning Codes, 2014; Property Topics and Concepts, 2007). Where specifications exceed a formula on one site – for example, a rule about the percentage of a site that can be nonpervious (unable to absorb rainwater) – it could affect specifications on a neighboring site.

The fourth, "stealth zoning" (a term we came up with, although making no claim we are the first to use it), involves the provision of flexibility on certain requirements if appearances, and outputs, are not exceeded. As long as the reasons behind the requirement are met, the smart code will allow flexibility. As an example, the smart code might allow more unrelated individuals to live together in a building if this is matched by an increase in bathrooms and exterior wall sound-proofing.

ICTs can help make these four principles of smart zoning feasible. ICTs can provide for real-time monitoring to adjust and adapt to situations on the ground. The code could allow different noise levels at different times per day, or provide a formula that would analyze data on different properties to adjust requirements such as setbacks or impervious surfaces. Algorithms implemented by ICTs – with their greater adaptability and flexibility based on the four principles of smart zoning – allow for a greater variety of streetscapes than uniform sets of regulations on items like setbacks and the number of unrelated individuals living in a building. Inflexible rules can lead new areas of construction to appear uniform and stale. However, the greater adaptability enabled by smart zoning can make for more interesting and variable neighborhoods, ones permitting more experimentation. These can be the neighborhoods attractive to young people, professionals and entrepreneurs, and members of the creative class. Such smart zoning techniques allow for greater adaptability to realities on the ground, making for more effective bylaws and regulations.

Barnett (n.d.) cites form-based coding to achieve urban designs which are more people centered. This is a move beyond broad-based engineering approaches which focus on square footage, parking, and traffic flows to the exclusion of walkability and street-level interactions. On the broader more directive-based forms of zoning, Barnett (n.d.) writes, "Not acknowledging the civic component of urbanism turns sidewalks and public spaces into utilitarian places between buildings, providing little more than light and air, and passages for pedestrians" (p. 3). A form-based approach would better account for this human element and, as shown in this chapter, ICTs can assist toward realizing this end. Monitoring and sensor technologies can better plan for and realize this human dimension in city planning. This will be shown in the discussion in this chapter and in the examples we employ.

Barnett (n.d.) highlights the role of technology, geographic information systems (GIS), as a powerful tool to "understand and describe natural systems at a variety of scales" (p. 2). The author states this can better inform the public and enable better decision making. The role of smart zoning in achieving this objective is a central consideration of ours as is the use of ICTs in achieving better decision making.

Smart regulation

The New Cities Foundation (2012) published a report on smart regulation, which does not deal as particularly with land use as smart zoning, but has many overlapping concepts and goals. Smart regulation also employs similar principles involving ICTs enabling new methods of governance that are more flexible and less rigid, allowing for adaptation in real time based on unique and changing conditions.

The New Cities Foundation identified three principles of smart regulation. First, embedment, which refers to the interrelation of different stakeholders – including private actors and local communities – and supporting their contribution to the design of neighborhoods. Second, responsiveness, which means regulations that respond to unique and changing conditions, through "multiple feedback loops" where stakeholders and users can react in real time when regulations fall short, calling on adjustments to be made (p. 22). Third, sustainability, which refers to regulations and policy aimed at promoting environmental sustainability and meeting social benchmarks such as equity. The sustainability principle involves the use of algorithms and ICTs to account for negative social and environmental externalities or effects (New Cities Foundation, 2012).

According to the New Cities Foundation (2012), ICTs that facilitated the involvement of multiple stakeholders and real-time feedback helped to "develop flexible polycentric systems that are able to respond more quickly to citizens' concerns" (pp. 28–29). This result is consistent and overlapping with the goals of smart zoning, because smart zoning also seeks to develop more flexible regulations and better real-time input.

Wikigovernment

ICTs can help with new modes of governance that provide greater room for public input and less for unilateral top-down decision making. In this context, wikinomics and wikigovernment are useful for considering a model of governance where there is greater feedback – feedback loops. Don Tapscott and Anthony Williams (2006) wrote in their book *Wikinomics* that while hierarchies have not disappeared, they have been challenged with ICTs and interactive Web 2.0. New democratic models are emerging "based on community collaboration, and self-organization rather than on hierarchy and control" (p. 1). With blogs and social media technologies, a multiplicity of voices provide information and feedback and empower people to speak up.

In *Macrowikinomics* (Tapscott and Williams, 2010), the follow-up to *Wikinomics*, the authors wrote that ICTs both create an environment with an active and collaborate citizenry and facilitate the "citizen collaborator" (p. 268). This creates a collaborative environment for governance where "today's governments need to distribute power broadly and leverage innovation, knowledge and value from the private sector and civil society" (pp. 263–264).

The examples in this chapter show how ICTs can increase transparency of government operations, more widely disseminating information and enabling new avenues of citizen collaboration and input. Smart zoning seeks an approach to land-use regulation that is less hierarchical, more flexible, and seeks input from citizens and unique situations on the ground to develop regulations that will be more nuanced and adjusted to different neighborhoods.

Sustainability and economic development

A goal of smart city technology is to foster economic growth and environmental sustainability. Robinson (2013) stated that "a smarter city creates sustainable, equitably distributed growth...with a focus on delivering social and environmental outcomes as well as economic growth" ("Define What a 'Smarter City' Means to You," para. 14). Blake (2013) writes that economic development and place making are connected, drawing on the thesis that dense, interesting, and mixed-use downtown-like neighborhoods are attractive to young professionals and entrepreneurs. Creating this street-level mix – bumping spaces where people randomly meet at coffee shops or on the street – fosters economic growth and development. Richard Florida (2013) has written about how urban San Francisco is gaining prominence over suburban Silicon Valley. For instance, San Francisco exceeds suburban Silicon Valley in venture capital investment. This shift is part of what Florida and others have termed "the great inversion" (Florida, 2013, para 5).

Caragliu, Del Bo, and Nijkamp (2011) wrote that "smart urban development" could be key in "fighting urban sprawl" (p. 69). Use of smart zoning algorithms can allow for land-use and building regulations that are flexible – regulations that provide a system of rewards and constraints where government acts as a conductor rather than a top-down social engineer. These would be in sync with citizens. Also, they could provide the dynamic people-centered neighborhoods, with walkable and bikeable streets well served by mass transit (transportability), that are conducive to creativity and entrepreneurial collaboration. This represents a confluence of digital urbanism and place making (Robinson, 2013).

One measure, the walk score, uses algorithms to measure walkability of different neighborhoods. Falk (2013) wrote of the walk score that it uses an algorithm to analyze the proximity (along walkable routes) of 10 million addresses to neighborhood amenities in 2,500 cities and 10,000 neighborhoods. An address is given a score up to 100 depending on the proximity to grocery stores, restaurants, shops, and so forth.

This is an important tool for both measurement and dissemination of information. With walkability an increasingly desirable quality in neighborhoods, such a score can assist potential home buyers and renters. This could also serve as a government transparency tool, as cities – and by extension municipal government policies – can be assessed based on walkability. In the United States, New York City (at 87.6%), San Francisco (at 83.9%), and Boston (at 79.5%) score the highest (Falk, 2013).

Walkability can be extended to the broader context of transportability – accessibility and mobility without using a personal vehicle, through means such as biking, walking, and mass transit.

Note that while a city can use ICTs to foster those goals the authors consider important such as transportability or mixed-use downtown-like developments, another city can use ICTs to do just the opposite. ICTs – and all of smart zoning – can restrict density and promote single-use areas along the lines associated with "cookie-cutter" suburbs – by tracking down home-based businesses or having measurements to make sure density remains at a low level. Ultimately, ICTs are important because they give policy makers and governments more choices.

Criticisms and sources of resistance to smart zoning

A city thinking about smart zoning should consider criticisms of smart zoning technologies. A major worry is privacy: the sensors and data collection associated with smart zoning technology can infringe privacy. However, there are ways to mitigate this. Noise sensors can be placed outside the boundaries of an individual site and designed to only monitor decibel levels, not the content of the noise (i.e., not monitor what is said in a conversation or played on a machine, but just how loud it is). Another argument is the cost of smart zoning technologies, though one can counter that such technologies contribute to economic development and to fostering an entrepreneurial climate, and are overall a fiscal plus.

Greenfield (2013) raised concerns about the algorithms associated with smart city technologies, stating that such algorithms fail to capture the complex nature of cities, the human element, and that algorithms reduce this complexity to simplistic unitary goals. Greenfield also raised concern that factors such as politics hinder the optimal solutions that smart city technologies purport to achieve. "We should all know by now," writes Greenfield, "that there are and can be no Pareto-optimal solutions for any systems as complex as a city" (para 14). He also raises concerns about democratic accountability, noting that none of the smart city literature suggests that the former would regularly apply to algorithms or algorithm designers (Greenfield, 2013, para 18).

A counter to charges that smart city (and smart zoning) technologies undermine democratic accountability is that ICTs can be a means to facilitate greater democratic engagement. Social media and Internet technologies provide new avenues for citizens to express their views, often directly to public officials who use outlets such as Twitter and

Facebook. Online technologies provide means to disseminate information, potentially increasing transparency and accountability.

When discussing the charge that algorithms oversimplify the realities of city life, Ward-Proud (2013) highlighted the need to avoid "overmanagement" by algorithms. However, if employed, ICTs and smart zoning can better account for the diverse realities of cities. Traditional fixed regulations lack the flexibility and adaptability of smart zoning techniques, and adding city-granted variances has historically not done a good job fixing the mismatches. Smart zoning can provide for greater feedback from situations on the ground, greater adjustability, and can facilitate greater variance in buildings and properties. Thus, smart zoning technologies can better capture the diversity and variability of cities.

The goal should be the appropriate amount of variability. A fixed rule that allows for city-granted variances can produce far more variety than most consider appropriate. The authors have seen numerous examples of residential assessment where a formula-based approach produces less variability and fewer cases of valuations that seem much too high or much too low.

Contrasting approaches to zoning – Asia and Latin America

In examining international examples of zoning, we found instances of top-down control where there was little citizen input. We also found instances of no zoning where there was no or little regulation or control. Our proposition is that smart zoning can offer a middle ground, with not only needed controls and regulations but also facilitating citizen input and responsiveness to situations on the ground. This will be seen in the examples we provide in the next subsections, but first, examples of top-down and nonexistent zoning. These international examples are not meant to be an exhaustive examination. Rather, they highlight some different approaches and how smart zoning relates to them. The discussion below shows how – in some of these countries – ICTs are being employed to counter the shortcomings of their current zoning system.

In much of Latin America, especially in the sprawling settlements around major cities, there was little or no effective land use or zoning (Lungo, 2001). So zoning was not only "not smart" but it was also "nonexistent." For these cities, the change cannot be from "nonsmart" to "smart zoning," but rather a complete starting from scratch. This does not preclude smart zoning from being implemented in these countries, but it shows that the method of implementation would differ from in North America.

In many Asian cities, Pacific Rim and the Indian subcontinent, land-use planning has been top down, often starting at the national level with little or no provision for public input (Goto, 1999; Gurran, 2011; Riethmüller et al., 1996). Furthermore, rules and procedures have often been ignored by national governments in implementing zoning policies (Riethmüller et al., 1996). However, several countries, including Thailand, Sri Lanka, and China, have experimented with more participatory approaches, with special attention given to involving people in the local area. Mostly, these efforts have used very traditional face-to-face meetings with little or no computer technology involved.

Some major cities in Asia, including Hong Kong, Singapore, and Kuala Lumpur, have tried for mixed-use zoning in the face of tremendous growth by using a top-down approach with zoning and planning (Teriman and Yigitcanlar, 2009). We have no evidence these cities have yet included smart technology to produce smart zoning, and they appear to have made little effort for the bottom-up citizen involvement often sought in the United States. We have evidence in Hong Kong of data gathering using smart technology, a key component of smart zoning, where an international group of scientists is studying air quality in three dimensions in the city (Cheung and Kao, 2014). If the study produces the data sought, it will lay the groundwork for a new dimension in performance zoning, because the city could monitor and then regulate pollution by altitude, latitude, and longitude.

There have been experiments with applying smart technologies, notably GIS, including in Nepal, Indonesia, Philippines, Sri Lanka, and Malaysia. As we would suspect, the problems were not with hardware or software, but rather with matching the technology to institutions and functional organizational setups.

Canadian experiments with smart zoning (or close analogies)

We have yet to find more than a few experiments of the sort we recommend to combine smart technologies with zoning to produce smart zoning. What we offer below, in this section and the next, are to show examples of what we recommend, not to demonstrate that our recommendations are already being carried out throughout the world. We do not know that, and we suspect the world has yet to see many experiments along the lines we suggest.

Fredericton – smart traffic control

Fredericton is a small city, with a population of 56,000 people, in the Saint John River Valley in the Canadian province of New Brunswick. It is the provincial capital, and home to two universities and an arts college. Fredericton has many elements associated with a strong creative economy – postsecondary institutions, strong arts and cultural sector, government, and a nearby military base in Oromocto. Its municipal government has emphasized a smart city strategy, making use of ICTs to improve service delivery and public administration. While it is hard to find examples of smart zoning, on traffic control there are strong examples of ICTs being used to develop traffic policy. Often, these traffic policy decisions overlap areas of zoning and land use as we discuss below. Overall, using ICTs – including monitoring technologies – in traffic control in Fredericton offers examples that could be applied more broadly to zoning, land use, and building design.

Smart monitoring technologies inform decisions on traffic policy in Fredericton. If residents of a neighborhood request traffic-calming measures, tubes placed on the street are deployed to measure traffic speed and see whether such calming measures – speed bumps, for example – are needed. The traffic tubes are relatively innocuous – most motorists do not know what they measure – and can collect information on the number of vehicles passing, the types of vehicles (e.g., trucks or personal vehicles), and speed. This determines whether the problem is serious enough to warrant traffic-calming measures. For instance, if the speed on a residential street is 60 kph or less, traffic calming is usually not carried out. While this is a traffic policy decision, it is one closely related to zoning. The relation is most notable when considering the need for traffic calming in certain neighborhoods based, in part, on zoning in such neighborhoods, and when using monitoring and measuring through ICTs to arrive at conclusions about the need for traffic-calming measures.

Another system of traffic monitoring used in Fredericton is Miovision, a camera system attached to a utility pole to provide video recordings of traffic at a select intersection, monitoring the number of vehicles. This video clip is uploaded to the Miovision platform and processed into traffic data, which is used to determine the traffic in the monitored area (Miovision, 2013). These data are used to determine the necessity of traffic signals and – closely related to zoning and land use – to inform developers about construction possibilities. More particularly, for residential areas the demand is for less traffic and quieter streets, while for commercial areas it is more traffic and more visibility. Monitoring for volumes of traffic can inform zoning and building decisions.

Smart city traffic technologies also provide tools to enhance transparency and dissemination of information to the general public. For a proposed roundabout, VISSIM software was employed to provide a three-dimensional simulation of pedestrian and motorized traffic on the roundabout, with existing architectural features of the neighborhood visible in the simulation, and cars and people. The data to make this simulation were developed by monitoring existing traffic and imposing the resulting data onto the roundabout design (City of Fredericton, 2013).

This visual simulation provides a valuable tool for policy makers, in being able to more accurately see the effects of a proposed change in roads. It serves an excellent public education function: the visual simulation of the proposed roundabout was shown at a public meeting and posted on the City of Fredericton's Facebook page, providing a visualization of its pedestrian and vehicular traffic readily accessible to residents.

Smart traffic technology has also been used in Fredericton as a measure to encourage greater compliance with the law. The City of Fredericton has installed speed radar signs at school zones which display traffic speeds by the roadside, encouraging people to slow down. A two-part study from the University of New Brunswick showed these display signs to be very effective in reducing speeds (Mason, 2010; Mason and Hildebrand, 2010; Paradis, 2011). While the smart zoning applications discussed in this chapter do not deal with behavior-changing uses of ICTs, the speed signs in Fredericton are an interesting example of smart technology use.

Saint John – ZoneSJ, citizen input, and transparency

Saint John is another New Brunswick city. According to the 2011 Canadian census, it has a population of just over 70,000 with a metropolitan (Census Metropolitan Area) population of 127,761. Saint John's economy is traditionally industrial, being the site of the Irving oil refinery and the Irving pulp and paper plant. For four decades, the city had seen a decline in population. However, the most recent Canadian census, 2011, showed growth for the first time in four decades, with Saint John's city center neighborhood – Uptown – becoming popular (Martin Prosperity Institute, 2013). In recent years, the City of Saint John has embarked on a comprehensive plan – Plan Saint John (PlanSJ) – which emphasizes centralized development, promoting the city's walkable and mixed-use Uptown.

Saint John – while not offering smart zoning in the strict sense of the term – provides excellent examples of wikigovernment in ICT applications to zoning and land use. One example is the use of ICTs to facilitate greater transparency in municipal government operations concerning zoning and land use and facilitating citizen engagement.

The City of Saint John's ZoneSJ (Zone Saint John) website offers an interactive zoning map which serves a public information and engagement function in the city's zoning process, where residents can see how zoning changes affect their neighborhoods. The map (available at http://maps.saintjohn.ca/zonesj-en) offers a satellite shot of the city overlaid with the zoning designations. This includes use (i.e., residential, commercial, mixed) and density (low rise, mid-rise, high rise) (ZoneSJ, 2013). By hovering over a zoning classification, greater details are provided, including a short explanation and a link to bylaw provisions.

Using ICTs to promote transparency and citizen engagement follows the aims of PlanSJ, as stated in a City of Saint John (2013) document:

> The successful implementation of PlanSJ will require collaboration and support from all members of the community…most importantly, it [PlanSJ] requires community partnerships and sustained engagement of the citizenry of Saint John to create the momentum of positive change needed to transform the City of Saint John. (p. 2)

Using web-based ICTs has enabled the City of Saint John – and the PlanSJ initiative – to better achieve these goals, providing new avenues to disseminate information and facilitate public input.

Another example of transparency and communication in Saint John comes from the city's mayor, Mel Norton, who posted the above-mentioned interactive zoning map on his Facebook page stating that "Your property is now on hi-tech zoning map www.saintjohn/zonesj thanks to world-class staff Yves Léger and Stacy Forfar" (Norton, 2013). Social media outlets, such as Facebook and Twitter, can serve as tools to further facilitate dissemination of information and citizen engagement.

The City of Saint John's website (http://www.saintjohn.ca/en/home/living/maps/default.aspx) offers air photos of the city and several interactive maps offering information on a range of services and policy areas, providing further avenues to inform residents. The site includes maps where users can obtain information on property and city services based on street address. These interactive maps provide information on a range of services including water management, heritage conservation properties, and solid waste and compost collection. All these help in dissemination of information to residents and in making the city's operations more transparent.

Toronto – online transparency

Saint John is not the only city to offer such interactive mapping tools. The City of Toronto (2013) – Canada's largest city – offers a similar feature, an interactive zoning map to help residents see the implications of the zoning bylaw and how it affects their neighborhoods (available at http://www.toronto.ca/zoning/schedule.htm). Much like Saint John, it offers a zoom-in/zoom-out map feature overlaid with zoning designations of property by use and density. Also like the Saint John map, Toronto's map provides a link giving more information on the particular zoning classification. However, the Toronto map offers only the bylaw provisions without a short explanation, and is somewhat less user friendly than the Saint John map.

Data and the Ontario Greenbelt

The Ontario Greenbelt is a conservation zone surrounding the Greater Toronto Area and the urbanizing Golden Horseshoe region along western Lake Ontario. The aim of the Greenbelt is to preserve forests and farmland and to curb sprawl, promoting intensification by designating a growth boundary around a fast-growing metropolitan region. Designating agricultural lands in the Ontario Greenbelt provides an example of algorithms in designing land-use policy. In identifying agricultural areas, two components are employed, Land Evaluation (LE) and Area Review (AR), which form the LEAR analysis. The LE component assesses capability of land for agriculture based on the Canada Land Inventory (CLI). AR assesses factors such as parcel size, fragmentation, infrastructure, and economic activity. Based on LEAR scores, each parcel is analyzed and given a score weighted for each of the LE and AR factors. By using GIS and LEAR analysis, lands are designated "agricultural" in Greenbelt plans (Ontario Ministry of Municipal Affairs and Housing, 2013).

This example shows how scoring and equations analyzed by a computer program can assist in the zone designation process, in this case conservation of agricultural lands from urban/suburban development.

U.S. experiments with smart zoning (or close analogues)

Again, we do not offer the examples below as proof that a widespread group of experiments are taking place. Instead, they are more examples to illustrate what we recommend should take place.

Austin, Texas

The city of Austin completed a comprehensive new city plan in early 2013, and is now engaged in a city-wide effort to update its land-use control code to follow the new plan. The effort, labeled "CodeNext," has been innovative in soliciting citizen input both in general and on a neighborhood-by-neighborhood basis. Conversations with city officials and their consultants, however, reveal that much of the thinking is *not* including what we call "smart zoning," or "IT-supported zoning." Little or no attention has been given so far to designing land-use controls that could be implemented or enforced with IT support.

We suspect, from our conversations with the city officials, consultants, and those attending the citizen input meetings, that at least two factors are at work. One, the city of Austin prides itself on "being weird," which means an individual-by-individual approach to city issues whenever possible. CodeNext has used the preexisting system of designated neighborhoods, each of which is only a few dozen blocks in size, and is seeking to identify and support the unique advantages of each. Algorithms, formulas, and other city-wide automatic activities are seen as a danger to preserving the uniqueness of each neighborhood.

Two, the Texas-wide tradition of "live and let live" is perhaps even more supported in Austin than in the rest of the state. In part, that means tolerance for many "violations" so long as no one complains and is found to have a valid complaint. Automatic enforcement of laws, including land-use controls, is often seen as not consistent with the Austin way of doing things.

We include Austin in part not only because one of us lives in the city, but also because the old code has, and the new code is likely to include, two provisions that could be implemented and enforced with IT support if the political will were there. The first is the noise ordinance, which specifies decibel levels for sound output. One could imagine placing sensors around the city to monitor noises, or even authorizing citizens with a smartphone app to perform the monitoring. Instead, enforcement occurs only when and if neighbors complain.

An objective means to monitor and report noise – sensors – could remove those awkward moments that can sour relations between neighbors should one of them call the police to make a noise complaint. A hurdle – having to make the complaint, call the police – might be removed, making noise bylaw enforcement easier and less "personal." This could also protect against frivolous noise complaints and ones that can harass law-abiding neighbors, when an objective decibel threshold is measured and must be met before law enforcement is deployed.

The second is the impervious area ratio, which sets standards for what percentage of a lot can be covered to be "impervious" to water seeping through. Because areas under decks, carport roofs, and the like are deemed "impervious," along with driveways, walkways, and other paved areas, one could imagine using satellite maps to calculate actual coverage and identify violations. Here again, such "automatic" enforcement is considered unacceptable, and the standards are applied when the owner makes a change requiring a permit and applies for that permit. Even then, the city relies on dimensioned drawings submitted by the developer, and only does its own measurement at final inspection.

So while one could imagine many requirements that could be implemented and enforced with IT support, such as lights, signage, exterior building materials, and others, we suspect the city of Austin will keep with its lot-by-lot procedures for political reasons.

Scottsdale, Arizona

We include Scottsdale because it has, in contrast to Austin, really embraced "citizens with smartphones" as "sensors" for a range of city problems. Citizens can use an app called "myScottsdale" (http://www.scottsdaleaz.gov/mediacenter/myScottsdale) to couple photos and GPS coordinates with the issue they seek to raise with the city. Incidentally, nothing appears to limit the app to actual citizens of Scottsdale. It is available on Google Play for Android, and presumably the Appstore for iOS, allowing people worldwide to download it. We wonder whether the GPS coordinates would limit its application to the Scottsdale city limits. The description we read (Violino, 2014) did not mention zoning/land-use issues alone, but presumably it could report code violations and potholes.

Using citizens with smartphones as sensors is important. One, it forestalls major investments in sensors by cities, especially because both the ratio of phones to the population and of smartphones to "devices less than smartphones" are growing rapidly. Approximately 90 percent of the population of a city is likely to have a cell phone, and that cell phone is now more often than not likely to have enough computing power (be "smart enough") to run apps like myScottsdale. Two, it changes the political dynamic in ways we find fascinating. Will this be seen as "citizens helping out" or as "citizens ratting out each other"? Given the world experience with spam constituting up to 90 percent of e-mail, what percentage of the submissions will be "valid" in the sense of pointing out a real issue? How about valid in the sense of coming from an actual citizen (and not tourists or visitors, which might be acceptable, but from malefactors or automated systems just trying to gum up the works)?

Citizens as sensors can apply to a lot more than land-use controls, but we include it here because it can play such an important role with those controls.

Seattle, Washington

Seattle's High Performance Buildings Pilot Project is *not* a zoning project per se. It is far more intrusive than most zoning because it involves an active partnership with the building owners to attach sensors inside each building's HVAC system to track and improve energy use. However, we include it in this chapter because it raises two issues we think may be key to smart (i.e., IT-supported) zoning.

One, it involves cooperation with the building owners in a win–win situation. The owners work with the sensors and the city because between them they can better monitor and control the energy use of the building. One could easily imagine such partnerships over many zoning-type issues – light, noise, emissions, vehicle and person visits, and so forth. The city would allow the use, subject to continuous monitoring; the owner would allow the continuous monitoring because that would give the owner better control over items crucial to its own operations.

Two, the data involved in such continuous monitoring mount up quickly. The Seattle project is a partnership with Microsoft and its cloud services because the project requires gathering, storing, analyzing, and reporting on such large bodies of data. In this pilot project, all the parties – owners, the city, and Microsoft – are hoping to acquire knowledge, skills, and capabilities that each can then apply in other ways: to other buildings, to other aspects of building operation, to other cities (City of Seattle, 2013).

Fitchburg, Wisconsin

Fitchburg is one of several cities that have adopted a "Smartcode District." We include it in this chapter not because it explicitly involves IT-supported "smart zoning," but because it is a more widespread movement (see Center for Applied Transect Studies, 2009) than our more focused definition of smart zoning, and it lays the groundwork for what could become smart zoning in our more focused sense (City of Fitchburg, 2012, 2014).

New York City

In New York City, the number one complaint on its 311 system is noise, something rarely tracked or handled, zoning or otherwise, although some uses are regulated on the expectation they will be noisy (Katz and Bradley, 2013, p. 29).

ICTs offer new opportunities. Noise can be tracked by modern smart sensors. One can envision a system of "quiet" and "less quiet" zones monitored by such sensors.

The municipal government of New York City is supporting an effort at Columbia University to develop a noise-tracking system that would enable "noise zoning" on not just the when, where, and how loud, but also potential source identifications.

Conclusion

We failed to find a single city that has implemented what we call smart zoning. Instead, we found a few places where the city is experimenting with land-use controls that either fully or at least partially meet our definition. We have presented here examples where cities have employed ICTs – or processes easily amenable to ICT use – to expand the parameters of zoning and building regulation, to provide innovative solutions in a context that at least partially meets the definition of smart zoning we have set out in this chapter.

Examples like Fredericton's use of ICTs for traffic monitoring and policy show the potential for such technologies in the zoning process, especially where there is overlap: where data on traffic patterns are used to determine the best locations for residential or commercial uses. ICTs also open up new avenues for citizen engagement and government transparency, as shown in the examples from Toronto and Saint John. Besides this, algorithms can optimize land use, as seen with the Ontario Greenbelt.

In the American examples, there are similar uses which – while not strictly smart zoning – are strongly analogous and can show the benefits of applying ICTs to the zoning process. In Austin, the desire for variance in neighborhoods could be enhanced through smart zoning algorithms. Scottsdale provides an instance of facilitating citizen activism through smartphone use. HVAC in Seattle shows how ICTs can adjust in real time to conditions on the ground, controlling energy use in a building.

The cities see not only the potential in their experiments but also recognize problems that will keep the implementation process moving slowly, rather than quickly. In a spirit of optimism, we summarize the problems first, and then conclude with the potential.

Problems with Smart Zoning

We discovered two sets of problems – technical and political. The two biggest technical problems were (a) the investment is up front; the returns are downstream; and (b) smart zoning requires new skills and resources. The two biggest political problems were (a) citizens have some inherent resistance to "mechanical" application and enforcement; and (b) those involved in land use, both administrative and elected officials, have little experience in encouraging and dealing with citizens, other than in one-site-at-a-time applications for a variance.

Technical Problems with Smart Zoning

The problem of up-front investment followed by downstream returns is well known to city governments, but usually when undertaking infrastructure investments, like roads and bridges, not in establishing land-use controls. Cities are used to land-use controls showing the opposite expenditure pattern: fairly cheap to draft and adopt, then expensive to implement with hearings, variances, and enforcement. Road departments have capital budgets; land-use departments do not.

The experiments so far show that smart zoning can require new technology (sensors, wiki pages), processing large amounts of data (not just sensor data, but also many more comments than land-use departments are used to receiving), and new skills in analyzing the data and making use of it. Again, the skills are not just for handling the technical data, but also for handling the increased citizen input.

Political Problems with Smart Zoning

Many citizens appear to have an inherent resistance to "mechanical" land-use controls, especially if they have yet to see any individual benefits from them. So using sensors, satellite views, and the like to identify and enforce land-use controls face the same resistance as using automatic speed detectors to enforce speed limits and traffic lights. The process seems "arbitrary" and "unfair" compared to relying on individual inspectors and police patrol cars.

Even if the city can demonstrate that it is less arbitrary and fairer, the resistance remains, in part because such processes are apt to catch a lot more people than the human process. The resistance will only subside when the citizens have more experience with such processes, and some of the downstream benefits (fewer disputes, lower enforcement budgets) emerge.

Note that the legal challenges that stem from the political opposition can have a very different basis if the sensor or other monitoring device is not on or over the property. To enter the property for inspection, or to place a monitor there, is subject to several regulations. To make observations via human or machine from a spot not on the property faces far less regulation. To fly over the property, via satellite, airplane, helicopter, or drone, is an emerging area of law currently unsettled (Molko, 2013).

Similarly, the zoning process is used to proceed with little citizen input, except for the rare cases where a request for a variance brings out largely local opposition. City officials are not used to requesting broader input, not just on decisions, but also on overall directions, especially if those requests turn out to be effective. They are also not used to dealing with the input they receive, lacking skills and technology in receiving it, analyzing it, re-presenting it back to the citizens, and taking action based explicitly on that input.

Technologies to help with that – web surveys, wiki pages, schemes to assign ratings and priorities to items, methods of analysis and presentation – are emerging, but are reasonably new and untested. In particular, cities have little experience with what are reasonable standards for how to weigh citizen input versus that of city officials versus that of subject experts. We are confident that if the cities keep experimenting, taking advantage of help from other cities, technical groups like Code for America, and land-use experts, cities will get at least as good at dealing with all this as they have with the site-by-site variance processes. We hope and expect that eventually cities will get even better outcomes (faster, cheaper, more accepted by citizens, and more beneficial to them).

Potential Benefits of Smart Zoning

Benefits remain in the potential stage because the cities are just experimenting with using IT support to make their zoning "smart" by our definition. However, we can identify four of what we think are some of the most important potential benefits.

Flexibility

Although the standard with variances system flexed when a site developer asked for it, the system was not flexible in advance. Using algorithms and output measurement via sensors and the like really provides for rapid, in-advance flexibility while preserving standards such as noise levels, emissions, pervious ratio, and the like.

Lower Cost

We predict the final return on investment in smart zoning will be high. It will take a few years for the lowered operational costs to surpass the initial investment, but it looks like the lowered costs will continue for decades into the future.

Better Understood and Accepted by Citizens

Soliciting, receiving, analyzing, presenting, and using citizen input is a lot of work. However, the end result, in the situations we have learned about, is that most citizens understand the policies better and accept them more than in cities that do not go through such a citizen involvement process. It is hard to put a number value on this, but informed and accepting citizens are key to a variety of community development, in economic terms and otherwise.

More Attuned to Land-Use Goals

The tools of smart zoning are powerful and appear to be more effective than the "fixed standards with variances available" system used in most cities. They can be used for goals the authors support; they can be used for goals the authors think are ill advised. The point is, the flexibility and lower cost enables a city to fine-tune smart zoning to better meet its chosen goals.

Recommendation

On balance, we think the potential benefits of smart zoning vastly outweigh its identified problems, and we urge every city to consider starting or continuing experiments in applying the techniques of smart zoning. The examples in this chapter, while not smart zoning per se, show the benefits of ICTs in ensuring more flexible applications of policy and in facilitating public input. The example from Fredericton closely overlaps into smart zoning, even if it is smart traffic management. There are many benefits for cities in embarking on this new frontier in public policy, including better zoning, planning, and public input. It is a direction worth taking.

References

Agricultural Marketing Service (AMS), U.S. Department of Agriculture (USDA), (n.d.). "Food deserts." Retrieved from http://apps.ams.usda.gov/fooddeserts/fooddeserts.aspx on January 9, 2015.

American Nutrition Association (ANA) (2010). "USDA defines food deserts," *Nutrition Digest*, 37(2). Retrieved from http://americannutritionassociation.org/newsletter/usda-defines-food-deserts on January 9, 2015.

Amherst Planning Department. (2012, March). "Innovative zoning in Amherst – a long history." *2012 Annual Town Meeting Planning Board Reports to Town Meeting and Supplementary Materials: Supplementary Information Regarding the Village Center Rezoning*. Retrieved from http://www.amherstma.gov/DocumentCenter/Home/View/18406 on February 24, 2014.

Barnett, J. (n.d.). "The way we were, the way we are: The theory and practice of designing cities since 1956." Retrieved from http://centercityphila.org/BIDclass/docs/10_barnett.pdf on January 30, 2015.

Beyer, P. (2012, August 1). "Overlay zoning districts." *Livable New York Resource Manual, II.2.h.* Retrieved from http://www.aging.ny.gov/LivableNY/ResourceManual/PlanningZoningAndDevelopment/II2h.pdf on February 24, 2014.

Blake, B. (2013, October 10). "Macon's College Hill Alliance: Change in local perception brings international recognition." *Knight Blog*. John S. and James L. Knight Foundation. Retrieved from http://www.knightfoundation.org/blogs/knightblog/2013/10/10/macons-college-hill-alliance-change-local-perception-brings-international-recognition/ on March 2, 2014.

Burchell, R. W., and Galley, C. C. (n.d.). "Inclusionary zoning: Pros and cons." *Inclusionary zoning: A viable solution to the affordable housing crisis?* Retrieved from http://ginsler.com/sites/ginsler/files/NHC-2.html on February 24, 2014.

Burns, P. K. (2014, January 14). "For a safer, healthier city, start walking." WYPR 88.1 FM News. Retrieved from http://news.wypr.org/post/safer-healthier-city-start-walking on February 24, 2014.

Caragliu, A., Del Bo, C., and Nijkamp, P. (2011). "Smart cities in Europe." *Journal of Urban Technology*, 18(2), pp. 65–82.

Center for Applied Transect Studies (2009). "SmartCode version 9.2." *The Town Paper*. Retrieved from http://www.transect.org/docs/3000-BookletSC-pdf.zip on February 24, 2014.

City of Fitchburg, Wisconsin (2012). SmartCode District Resources. Retrieved from http://www.city.fitchburg.wi.us/departments/cityHall/planning/SmartCodeResources.php on February 28, 2014.

City of Fitchburg, Wisconsin (2014, February 24). Fitchburg WI, Code of Ordinances. Municipal Code Corporation. Retrieved from http://www.municode.com/Library#/WI/Fitchburg on February 28, 2014.

City of Saint John (2014). Maps and Air Photos. *City of Saint John, Living*. Retrieved from http://www.saintjohn.ca/en/home/living/maps/default.aspx on February 28, 2014.

City of Saint John (2014). ZoneSJ. *City of Saint John, Growth & Community Development, Community Planning*. Retrieved from http://www.saintjohn.ca/en/home/cityhall/developmentgrowth/communityplanning/zonesj/default.aspx on February 28, 2014.

City of Scottsdale, Arizona (2014). myScottsdale. *Scottsdale Media Center*. Retrieved from http://www.scottsdaleaz.gov/mediacenter/myScottsdale on February 28, 2014.

City of Toront. (2013, May 9). Zoning Bylaw 569-2013 Zoning Map. Retrieved from http://map.toronto.ca/maps/map.jsp?app=ZBL_CONSULT on February 28, 2014.

City of Seattle (2013, July 9). Seattle Launches High-Performance Building Pilot. *Sustainable City Network*. Retrieved from http://www.sustainablecitynetwork.com/topic_channels/building_housing/article_d6c015e6-e8f4-11e2-a8f6-001a4bcf6878.html on February 28, 2014.

CodeNEXT (2013, December 12). Survey Runs through December + Community Character in January. Imagine Austin. Retrieved from http://www.austintexas.gov/blog/survey-runs-through-december-community-character-january on February 24, 2014.

Cole, R., Long, M., and Cole, I. (2013, August 28). *Wholesale economic development. – I. approach.* [Kindle book] Retrieved from Amazon.com.

Concord Square Planning and Development, Inc. (n.d.). *Massachusetts General Laws Chapter 40R.* Retrieved from http://www.concordsqdev.com/home/smart-growth-zoning/ on February 24, 2014.

D.C. Office of Planning (June 2008). *Zoning Best Practices.* Retrieved from https://www.communicationsmgr.com/projects/1355/docs/Zoning_BPs_whole.pdf on February 24, 2014.

Electronic Frontier Foundation (2012). Surveillance Drones. Retrieved from https://www.eff.org/issues/surveillance-drones on February 28, 2014.

Falk, T. (2013). "Top 10 Most Walkable U.S. Cities." *The Bulletin,* 11 (November 8). SmartPlanet.com. Retrieved from http://www.smartplanet.com/blog/bulletin/top-10-most-walkable-us-cities/33747?tag=content;siu-container on February 24, 2014.

Florida, R. (2013, August 5). "Why San Francisco may be the new Silicon Valley." *The Atlantic Cities.* Washington, DC: The Atlantic. Retrieved from http://www.theatlanticcities.com/jobs-and-economy/2013/08/why-san-francisco-may-be-new-silicon-valley/6295/ on March 2, 2014.

Goto, T. (1999). "Land use control regulation in Japan." Tokyo, Tokyo Conference of the Asian City Development Strategy. Retrieved from http://www.gdrc.org/uem/observatory/land-regulation.html on January 15, 2015.

Green Valley Institute (n.d.). Innovative Zoning Techniques. Retrieved from http://www.greenvalleyinstitute.org/landuse_innovativezoning.htm on February 24, 2014.

Greenfield, A. (2013). "Against the smart city." *Urban Omnibus,* October 23. The Architectural League. Retrieved from http://urbanomnibus.net/2013/10/against-the-smart-city/ on February 24, 2014.

Gurran, N. (2011). *Australian Urban Land Use Planning: Principles, Systems, and Practice: Second Edition.* Sydney, Australia: Sydney University Press.

Hirt, S. (2007, Autumn). "The devil is in the definitions: Contrasting American and German approaches to zoning." *Journal of the American Planning Association,* 73(4). Retrieved from http://www.archive.spia.vt.edu/SPIA/docs/shirt/The_Devil_is_in_the_Definitions.pdf on January 9, 2015.

Holeywell, R. (2013, December 13). "The most walkable cities and how some are making strides." *Governing.* Retrieved from http://www.governing.com/topics/urban/gov-most-walkable-cities.html on February 24, 2014.

Housing Innovations Program (2014). Tool: Incentive Zoning. Puget Sound Regional Council. Retrieved from http://www.psrc.org/growth/hip/alltools/incent-zoning/ on February 24, 2014.

Inniss, L. B. (2007–2008). "Back to the future: Is form-based code an efficacious tool for shaping modern civic life?" *University of Pennsylvania Journal of Law and Social Change,* 11, pp. 75–103. Retrieved from https://www.law.upenn.edu/journals/jlasc/articles/volume11/issue1/Inniss11U.Pa.J.L.&Soc.Change75%282007%29.pdf on January 9, 2015.

Iverson, C. (2013, November 18). "Time to change Minneapolis zoning codes." *Minnesota Daily*. Retrieved from http://www.mndaily.com/opinion/columns/2013/11/17/time-change-minneapolis-zoning-codes on February 24, 2014.

KaBoom (2015). "What is a play desert?" Retrieved from http://kaboom.org/map_play/what_play_desert on January 9, 2015.

Karkkainen, B. C. (1994). "Zoning: A reply to the critics." *Journal of Land Use & Environmental Law,* 101. Retrieved from http://www.law.fsu.edu/journals/landuse/vol101/karkkain.html on January 9, 2015.

Katz, B., and Bradley, J. (2013). *The metropolitan revolution: How cities and metros are fixing our broken politics and fragile economy.* Washington, DC: The Brookings Institution.

Lungo, M. (2001, March). "Urban sprawl and land regulation in Latin America." *Land Lines,* 13(2). Retrieved from http://www.lincolninst.edu/pubs/255_Urban-Sprawl-and-Land-Regulation-in-Latin-America on January 15, 2015.

Martin Prosperity Institute (2013, February 12). "Insight: Even smaller cities are growing." Retrieved from http://martinprosperity.org/2013/02/12/insight-smaller-cities-growing/ on March 3, 2014.

Mason, First Name. (2010). "The effectiveness of speed display boards in reducing speeds in school-zones." Fredericton, New Brunswick, Canada: University of New Brunswick. (Term Paper CE 5943 – Research Project; Supervisor: Dr. Eric Hildebrand)

Mason, First Name, and Hildebrand, E. (2010). "The effectiveness of radar speed displays in school-zones: A Fredericton case study." Fredericton, New Brunswick, Canada: University of New Brunswick.

Miovision.com (2013). Scout Video Collection Unit. Retrieved from http://miovision.com/traffic-data-solution/scout-video-collection-unit/ on March 3, 2014.

Motoyama, Y., and Konczal, J. (2013). "How can I create my favorite state tanking?: The hidden pitfalls of statistical indexes." *Journal of Applied Research in Economic Development,* 10. Retrieved from http://journal.c2er.org/2013/09/how-can-i-create-my-favorite-state-ranking/ on February 24, 2014.

Mukherji, N., and Morales, A. (2010). "Zoning for urban agriculture." *Zoning Practice Issue,* 3(March), Practice Urban Agriculture, American Planning Association. Retrieved from http://www.planning.org/zoningpractice/2010/pdf/mar.pdf on February 24, 2014.

Nashua Regional Planning Commission (n.d.). "Performance zoning." *Integrating Transportation and Community Planning Fact Sheet 34*. Retrieved from http://www.nashuarpc.org/landuse/PDF_files/fact_sheets/FS34_Performance_Zoning.pdf on February 24, 2014.

National Recreation and Park Association (NRPA) (n.d.). "Play deserts." Retrieved from http://www.nrpa.org/Grants-and-Partners/Recreation-and-Health/Play-Deserts/ on January 9, 2015.

New Cities Foundation (2012). *Smart Regulation.* "Explaining Key Concepts for the Cities of Tomorrow" series, Discussion Paper #1. http://www.newcitiesfoundation.org. Retrieved from http://issuu.com/newcitiesfoundation/docs/new-cities-foundation-smart-regulation?e=5065274/1327018 on February 24, 2014.

Nicholson, C. (2014, February 17 and 24). "Q&A: Marcus Quigley on how the Internet of Things is automating city systems." *The Bulletin,* I11. SmartPlanet.com. Retrieved from http://www.smartplanet.com/blog/pure-genius/qa-marcus

-quigley-on-how-the-internet-of-things-is-automating-city-systems/?tag=nl.e662&s_cid=e662&ttag=e662&ftag=TRE383a915 on February 24, 2014.

Norton, M. (2013, November 2). "Your property now on hi-tech zoning map." [Facebook post] Retrieved from https://www.facebook.com/melnortonsj/posts/10151973678161181 on February 28, 2014.

Ontario Ministry of Municipal Affairs and Housing (2013). "Building a greenbelt." Land Use Planning, Greenbelt Protection. Retrieved from http://www.mah.gov.on.ca/Page1396.aspx on March 3, 2014.

Ottensmann, J. R. (1999–2000). "Market-based exchanges of rights within a system of performance zoning." *Planning & Markets,* 1(1).

Paradis, D. (2011). A Retrospective Analysis of Speed Reduction Strategies. CE 5943 Research Project Term Paper. University of New Brunswick.

Peirce, Neil. (2013, August 23). "Smart city thinking." *Citiwire.net,* Sunday, August 25, 2013. Washington Post Writers Group. Retrieved from http://citiwire.net/columns/smart-city-thinking/ on February 24, 2014.

Planning Division. (2014). City of Boise. Retrieved from http://pds.cityofboise.org/planning/ on February 24, 2014.

Pogodzinski, J. M. (1991). "The effects of fiscal and exclusionary zoning on household location: A critical review." *Journal of Housing Research,* 2(2), pp. 145–160. Retrieved from http://content.knowledgeplex.org/kp2/img/cache/kp/2566.pdf on January 9, 2015.

Riethmüller, R., and The Asian Working Group on Land Use Planning for the Asian-Pacific Region (1996). *Experiences of land use planning in Asian projects: Selected insights*, Colombo 1996, Eschborn, Germany, Deutsche Gesellschaft für Technische Zusammenarbeit (GTZ) GmbH, Division 425 Multisectoral Urban and Rural Development Programmes. Retrieved from https://www.mpl.ird.fr/crea/taller-colombia/FAO/AGLL/pdfdocs/lup-asia.pdf on January 15, 2015.

Robinson, R. (2013, September 8). "Seven steps to a smarter city: And the imperative for taking them." *The Urban Technologist* (blog). Retrieved from http://theurbantechnologist.com/2013/09/08/seven-steps-to-a-smarter-city-and-the-imperative-for-taking-them-updated-8th-september-2013/ on March 3, 2014.

Schleicher, D. (2013, May). "City unplanning." *The Yale Law Journal,* 122(7), pp. 1670–1737.

Smart City Memphis. (December 15, 2011). "We need more zoning." Retrieved from http://www.smartcitymemphis.com/2011/12/we-need-more-zoning/ on February 24, 2014.

SmartGrowth Online (2013, June 12). "Testing the results of municipal mixed-use zoning ordinances: A novel methodological approach." U.S. Environmental Protection Agency. Retrieved from http://www.smartgrowth.org/engine/index.php/resources/2013/06/12/testing-the-results-of-municipal on February 24, 2014.

SmartGrowth Online (2013, October 30). "Examples of codes that support smart growth development." U.S. Environmental Protection Agency. Retrieved from http://www.epa.gov/dced/codeexamples.htm on February 24, 2014.

Tapscott, D. (2013). "Introducing global solution networks." *Global Solution Networks*. Retrieved from http://gsnetworks.org/introducing-global-solution-networks/ on February 24, 2014.

Tapscott, D., and Williams, A. D. (2008). *Wikinomics: How mass collaboration changes everything* (expanded ed.). Portfolio Hardcover. ISBN-10: 1591841933.

Tapscott, D., and Williams, A. D. (2010). *Macrowikinomics: Rebooting business and the world.* Portfolio Hardcover. ISBN-10: 1591843561. ISBN-13: 978-1591843566.
U.S. Environmental Protection Agency.(2013, October 30). "Examples of codes that support smart growth development." Retrieved from http://www.epa.gov/smartgrowth/codeexamples.htm on February 24, 2014.
Violino, B. (2014, February 12). "Smart cities are here today – and getting smarter." *Computerworld.* Retrieved from http://www.computerworld.com/s/article/9246179/Smart_cities_are_here_today_and_getting_smarter?source=CTWNLE_nlt_thisweek_2014-02-17 on February 24, 2014.
Ward-Proud, L. (2013, October 30). "Smart cities have arrived and the potential is huge." *City A.M.* Retrieved from http://www.cityam.com/article/1383095607/smart-cities-have-arrived-and-potential-huge on February 24, 2014.
Wickersham, J. (2001). "Jane Jacob's critique of zoning: From Euclid to Portland and beyond." 28 B.C. Envtl. Aff. L. Rev. 547, Retrieved from http://lawdigitalcommons.bc.edu/ealr/vol28/iss4/5 on January 9, 2015.
Williams, F. B. (1922). "The law of city planning and zoning," in Ely, Richard T. (ed.), *The Citizen's Library of Economics, Politics and Sociology*. New York: Macmillan.

12
New Mobilities for Accessible Cities: Toward Scenarios for Seamless Journeys

Barbara Adkins, Marianella Chamorro-Koc, and Lisa Stafford

Abstract

While cities increasingly attest to plans to make their resources accessible for people with disabilities, the realities of achieving the travel considered integral to urban life continue to be frustrating and prohibitive for this group. Accessing the basic opportunities of contemporary urban life now presupposes the supports and resources afforded by new mobilities, combining virtual and actual travel and communication in negotiating our work, leisure, connections with families, and culture. For the researchers applying the new mobilities paradigm, this requires a focus that is suited to capturing movement and its spatial and temporal coordinates and should also turn to illuminate the darker side of these relationships: coerced immobility experienced by people with disabilities. This chapter discusses an approach to research and the development of design scenarios – concepts emerging from research that may inform design – that take seriously the role of movement, time, and space in the achievement of valued connections by individuals with disabilities with particular reference to the journey to work. In particular we apply, in a case study, concepts of time and space that are relevant to the in situ experience of getting to work; raising questions regarding the way getting ready and travelling are experienced in the context of risk and contingency, and the actual and potential role of the technical, material, and social environment. We then respond to the analysis of this case with a discussion about the way emergent scenarios can imagine "possible or preferable futures" for the mobile citizenship of people with disabilities.

Introduction: Democracy, the city, and the journey to work for people with disabilities

With reference to transport planners' broader concept of mobility as the individual freedom to move (Walker, 2012), we argue that there is a lot at stake in addressing the journey to work experienced by people with disabilities, and there is potential for design to play a significant role. A synthesis of the relationship between disability and work across Organisation for Economic Co-operation and Development (OECD) countries reports that employment rates for people with disabilities are 40 percent below the national level on average and unemployment rates are typically twice the overall level. Given the clear requirement for many people with disabilities to rely on benefits, these statistics also point to high social costs associated with these patterns including much lower incomes and higher poverty risks (OECD, 2010: 10). While this brief overview clearly points to access to work as central to enhanced social and economic participation, it is crucial that we understand the challenges of working for people with disabilities at the experiential level (Marston and Lantz, 2012). This chapter focuses on a key aspect of these challenges: the daily rhythms and routines of getting to work and the way in which the journeys are experienced symbolically, and thus have implications for identity and a sense of participation and citizenship. Drawing upon these insights, we raise the question of how the affordances of digital technologies may potentially be engaged in research and debate concerning the democratisation of the city in the field of working life. We develop a conceptual framework and examination of a case example that sheds light on the way technologies may be actually and potentially involved in the experience of getting to work on the part of people with disabilities at the level of daily experience. We first locate the question of the role of digital technologies in this domain through a conceptual review of the nature and evolution of our contemporary experience of the city with respect to the space–time relationships of mobility and automobility, arguing that design needs to focus on the agencies at stake in people–technology–environment relationships. We then report on a case study of the daily experiences of a person living with muscular dystrophy in his journey to work, drawing on interview data as well as daily diaries focusing on these experiences. We conclude with a discussion of the way the study of such experiences can inform scenarios that take seriously the full set of social–spatial–temporal–technical relationships that can transform the journey to work from an experience of "misfitting" (Garland-Thompson, 2011) into one that is consistent with mobile citizenship.

Conceptualizing the time and space of work in the city: Automobility and participation in working life

The rise of a "mobile society" has been the impetus for scholars to argue for an approach to understanding sociality that takes movement, flux, flow – and also immobilities – into account. This focus on mobilities requires us to bring urban space into view as the site of different and intersecting types of travel: corporeal travel of people, physical movement of objects, imaginative travel through narratives, virtual travel on the Internet, communicative travel in person-to-person messages and conversations (Urry, 2008). Our positioning in relation to these systems of mobility is seen as central to the formation and sustainment of the self: its everyday activities, interpersonal relations with others, as well as connections with the wider world (Urry, 2012: 6). Public transport professionals and transit planning designers have defined personal mobility as "a freedom," or the ease of moving about, for people or goods within the transportation system (Walker, 2012). However, planners also consider that personal mobility is not necessarily about movement, but about people's ability to access where they want to go (Walker, 2012: 19). Thus, our relationship to mobility is central to the way we are now positioned with respect to mobile citizenship (Cass, Shove, and Urry, 2005) that emerges as a key stake in access to contemporary cities.

For Urry (2008), a focus on mobilities entails understanding the emergence of the space–time relationships that characterise contemporary social life. Mobilities required for working life have emerged historically through the creation of social conditions of industrial production. For Marx, work under capitalism creates a temporality to which workers are subject. Time is commodified. Clock time becomes a measure of work which is disembedded from its social and physical context. In responding to the requirements of clock time, people require the application of disciplines associated with "being on time" and "putting in the required amount of time," subscribing to the importance of not wasting time, and of managing one's own time with diligence (Urry, 2008: 109).

In the context of these requirements to optimise the use of time, people may be differently resourced to do so. Certainly, having the economic resources to enable compliance with temporal constraints such as money, locational capital, technology, and social supports can make a difference to people's capacity for compliance and also to the qualities of time that are experienced. These are needed to address potential tensions between clock time and other aspects of sociality. For Barbara Adam (1995: 45), a key resource in the quality of time is associated with the body and well-being. For Adam, people's sociality and well-being have always been associated with biological and environmental rhythms where our body rhythms are synchronised with the rhythms of the environment (e.g., night and day, the seasons, and so on). Industrial time, however, is superimposed on these "nested body and planetary times" (Adam, 1995: 46), exemplified by clock time and artificial light. Clock time has been central to the shift in emphasis in everyday living and working patterns from "variable rhythms to invariant ones" (Adam, 1995: 47). Further, overlaid on these temporalities is a culture of speed and pace: "When time is money, faster is better" (Adam, 2006: 124). The tensions created by the requirement to embed in practice the constraints and disciplines of clock time become visible where people are now supported with time-slowing strategies for health and well-being such as meditation, biofeedback, and hypnosis. Issues of clock, time, and resources in everyday life activities are comprised in what we understand as "access." Litman (2011) discusses three forms of access; the first by traveling; the second by using telecommunications; and the third by relocating closer to the desired resources. In this view, mobility is one dimension of access (Walker, 2012).

In addition to the importance of speed and pace in our capacity to participate in work, the era of automobility has also heralded a greater expectation of flexibility and commuting practices that may require journeys of significant length. For Urry (2004: 26), "'auto' mobility involves autonomous humans combined with machines with capacity for autonomous movement along the paths, lanes, streets and routeways of one society after another." This expectation now extends beyond the use of automobiles themselves to more general culturally embedded assumptions regarding disciplines of time and space. Historically, the system of automobility led to the emergence of new assumptions regarding the spatiotemporal relationships involved in the workday, dividing "workplaces from homes, [and] producing lengthy commutes into and across the city" (p. 28). Further, while it currently provides greater "freedoms" enabling autonomous individuals to regulate their own journeys which may now appear more convenient and less fragmented than other systems of mobility, automobility also carries with it constraint and coercion. Assumptions of this flexibility and self-regulation come to be built into the conditions that govern entry to fields such as working life. Automobility comes to be embedded in our sense of normality.

There is much at stake in people's capacity to participate in this normality, such as the predictability and relative seamlessness of the journey to work. This normality and predictability is associated with seven phases of a trip when considering public transport: understanding of the service, accessing at origin, waiting, paying, riding, connecting, and accessing at destination (Walker, 2012). These constitute the expected pattern of a public transport journey. The analyses of Erving Goffman (1967, 1971) expose the importance of being able to trust in the patterned and predictable practices of everyday life, both in our capacity to maintain this for ourselves and to trust that others will also sustain these patterns (Misztal, 2001). The stake here is not simply the maintenance of social parameters and points of reference for the purpose of undertaking daily functions. Keeping chaos and unpredictability at bay is also central to the capacity of people to maintain a focus on self and identity – the capacity to "preserve something of oneself from the clutch of an institution" (Goffman, 1971: 319). Since Goffman's observations were published, the contemporary importance of this capacity to shape and adapt identity has been well documented. However, this capacity is threatened for many people with disabilities who are subject to "coerced immobility" (Hughes, Russell, and Paterson, 2005; Urry, 2002). They are disadvantaged in relation to keeping daily functions and activities under control, and also in terms of opportunities to participate in the shaping of identities.

Mobile citizenship, disability, and technology

For the purposes of this chapter, the potential for people with disabilities to be severely disadvantaged with respect to mobile citizenship raises important questions pertaining to the role of technologies in addressing this coerced immobility. However, the suitability of technologies and their uses for this purpose cannot be assumed. In relation to the role of digital and new media technologies in fostering participation more generally, Brighenti (2012: 409) asks, "Do new media operate primarily as bridges or walls? Are they more suitable for empowering or surveilling? They are tools—but for whom? They are environmental—but what kind of social environments and spaces do they create?" These questions need further refinement in relation to their role in supporting the mobile citizenship of people with disabilities. For Moser (2006), some assumptions underpinning the role of technologies in relation to people with disabilities rest on the notion that they are primarily oriented to "normalising" the disabled body. These assumptions with respect to technology are derived from a broader tendency in the positioning of people with disabilities across diverse fields of social life.

For example, in their study of young people with disabilities in the field of consumption, Hughes et al. (2005: 12) assert that they are positioned differently in the field because "The qualities associated with disability and impairment symbolise negative value and deficit with respect to physical, cultural and social capital." This can endanger their status as citizens because "the meaning of impairment is transformed from an attribute of a person to a master status that makes it absolutely equivalent to the anomaly. This process preserves the integrity of the 'normal' but objectifies, dehumanises and invalidates disability" (Hughes et al., 2005: 13). This marginalisation of disability is the outcome of policy, planning, and infrastructure assumptions, market constraints, as well as the significant layer of cultural assumptions and dispositions that can constitute prohibitions and barriers. Together these economic, spatial, technical, social, and cultural challenges can produced the coerced immobility described above in a society where new mobilities are central to citizenship. For Hughes et al. (2005: 14), this requires a more concerted and specific focus on "the nature and extent of 'coerced immobility' and how it impacts on particular social groups." Thus, research informing design that can address tendencies for coerced immobilities requires a focus on actual and virtual mobilities including the agencies that produce frustrated mobilities. Further, following Moser (2006), this requires an approach that is open to capturing the nature and distribution of agency across people, technologies, and environments and thus resists unhelpful assumptions that emanate from disability as a master status.

The focus of this chapter is on the relationship of people with disabilities to mobile citizenship and the agencies that are decisive in providing access to this citizenship, with a particular focus on the journey to work. This is considered an essential first step for producing emergent "scenarios" that are responsive to the way the distribution of agencies currently operate with a view to promoting mobile citizenship. Scenarios are conceptual frameworks that describe "possible, preferable or avoidable futures"; they are employed for disciplinary design development where the agencies described above are the centre of the design production. The design of scenarios is informed by a cyclical reflective process between theory and practice that leads towards a "prospective" design solution (Chamorro-Koc, Buccolo, and Adkins, 2012) grounded in an identification and understanding of use situations (Carroll, 1997: 385). The next section describes our approach to the case study we have selected to illuminate these agencies in relation to the mobilities and immobilities associated with the journey to work and reports on our findings.

Case study and findings

This section describes some of the everyday experiences of a person with muscular dystrophy[1] in his daily efforts to participate in working life. We will call him Louis. Louis's journey to work is clearly not necessarily representative of the diversity of experiences on the part of people with disabilities. He lives with a specific mobility-related disability, and is extremely committed to participating in working life as a case worker for a disability service provider. His commitment and determination is also manifested in other fields. He is currently training as a para-athlete, with a vision of competing internationally. Our case thus highlights some of the significant, but largely invisible, barriers to getting to work for people that share mobility-related disabilities, with the qualification that Louis has more resources, support, health, and fitness than many people living with these disabilities. We describe the experience of getting to and from work in a day in the life of this participant. To capture this lived experienced, this case study used participatory methods, such as self-diaries and visual probes, and qualitative interviews. These methods enable participants to step back and describe daily routines that can be quite habitual and unremarkable.

Our approach to the framing and analysis of the case is based on the emphasis on the part of some disability scholars on the spatiotemporal context in which access to the city is experienced (Freund, 2001; Gleeson, 1999; Kitchin, 1998), and in these settings, an understanding of the way this experience becomes embodied: the "cognitive-sensual dispositions towards spaces (the habitus) that develop, become stable, sedimented and 'second nature'" (Freund, 2001: 701). At this level of daily lived experience we also employed the concepts of normality and trust as employed in the analyses of Erving Goffman (1967, 1971), subsequently reviewed by Misztal (2001), in order to focus on the ordinary human and nonhuman agencies that together constitute productive, discouraging, or unsuccessful journeys to work. Further, we locate the study in the context of the social relationships entailed in public transit in light of key expectations that people articulate when considering using a transit service:

1 *It takes me where I want to go.*
2 *It takes me when I want to go.*

3 *It is a good use of my time.*
4 *It is a good use of my money.*
5 *It respects me in the level of safety, comfort, and amenity it provides.*
6 *I can trust it.*
7 *It gives me freedom to change my plans (Walker, 2012: 24).*

We focus on these relationships of transit because, at the time of collecting the data, Louis, like many people with physical disabilities, did not have access to an appropriate vehicle to drive, and was reliant on other forms of transport including public transport. However, his experiences with public transport do not fit with Walker's description of the seven phases of transit trip, and constitute experiences of "misfitting." His accounts point to the sheer amount of organizational effort that is necessary for achieving the journey to work in a way that is consistent with the requirements of clock time and work time and the disciplines required in a system of automobility. The accounts will also show that all this organizing is a necessary – but not sufficient - condition for getting to work successfully, illustrating the contingencies that can thwart or disrupt the journey in spite of Louis's best efforts. Further, we discuss the way that these issues not only threaten the daily rhythms of work but also represent constant challenges to his sense of place in the field of urban life. We now turn to discuss the insights from the case study in terms of the rituals and routines of the daily journey to work, including getting ready and the physical journey itself, the nature of the experience of the journeys, and the implications of these experiences for his sense of mobile citizenship.

Habitual routines: getting ready and travelling

For Louis, there are two distinct spatial rhythms located in the journey to work: getting ready and travelling. Each of the sequential body–time–space–technology interactions illuminates the amount and nature of agencies that are enacted routinely to sustain normality.

Getting Ready

Getting ready for work begins the night before, sometimes entailing the requirement to prebook the transport needed to get to work. Not every taxi can accommodate a wheelchair and in Brisbane, Australia, the site of our study, wheelchair-compatible buses cannot be assumed and booking is necessary. The evening ritual for Louis entails plugging in the devices that enable performing the role of the worker the next day: the mobile phone and the wheelchair. The wheelchair affords both movement and "reach." Reach is crucial in addition to wheeling. Many activities in everyday life – for example, our daily use of objects and devices such as door handles and locks – carry with them the assumptions of a standing position. He describes the agency of the wheelchair as follows:

> I have a wheelchair that has power-assisted wheels called emotion wheels. [It is] a levo wheelchair that enables me to stand and reach things you simply can't in a sitting position along with the health benefits you gain from being in a standing position.

In addition to its essential role in conducting his work the following day, the mobile phone is important for planning and organizing for the working day the evening before, assisting in organizing Louis's journey to work the following day and setting up and organizing his transport options and times. It also enables the setting of the alarm to ensure that there is enough time the following morning to wake up before the support worker arrives at 6.00 a.m., and to undertake activities in time to leave the house and go to work. In his diary he notes his evening ritual:

> Wheeled alongside my bed, changed my clothes, set the timer for 5.50 a.m., plugged in my phone and chair to be charged for the next day and transferred onto my bed. Goodnight.

The next morning starts with waking to the alarm. Then, in order to help Louis keep to the routine of getting to work on time, a support worker arrives at 6:00 a.m. Whilst he is getting ready by showering and getting dressed, the support worker helps by attending to daily activities that he could not complete himself in time for leaving for work, such as preparing lunch, tidying up the dishes, and ironing clothes. While this routine is relatively stable, it is important to recognise that the agencies entailed in Louis's own efforts and activities, the contribution of the support worker, and the technology are all required, and in a particular sequence, for the journey to work at an appropriate time to be possible. If there is a disruption at any point in the routine – for example, if there is a disruption to power (so that the wheelchair is not charged) or the personal carer is unable to come – the spatial-temporal relationship for automobility and work is in jeopardy. If all goes to plan, after having breakfast and cleaning his teeth, Louis has time to check e-mails before undertaking the next important routine: leaving home to go to work.

Travelling to work

The reliability of the second process, the act of travelling to work, is more tenuous, where circumstances that disrupt mobility are quite frequent. In the absence of the option of driving his own vehicle, Louis describes his travel options in terms of a commuter using public transport, a passenger in the vehicle of a coworker, or in an accessible taxi:

> travelling to and from work varies from being picked up, catching a bus and wheeling 15 minutes from the bus stop to work and of course the occasional cab but that can be pricey and adds up quickly.

Each of these options entails a specific spatial, temporal, and technological and social set of agencies. The journey in a vehicle driven by a coworker is the most successful journey, agreed upon the day or evening before the journey and entailing the least risk of disruption or immobility, due to enhanced flexibility and interpersonal understanding of the needs of the journey. However, in the role of commuter, in the taxi or the bus, in spite of Louis's organizational efforts, is very prone to contingency and difficulty. In relation to the taxi, which is ordered the previous evening, the relative scarcity of taxis that can accommodate wheelchairs can entail greater uncertainty concerning punctuality, and, indeed, whether it will arrive at all.

With respect to the bus, even when a wheelchair-accessible bus has been ordered, there is always significant risk. In spite of ordering a wheelchair-accessible bus, there is always a chance that the bus will not be wheelchair accessible. Sometimes, even when the bus is wheelchair accessible, there is the risk of not being seen if he is the only passenger at the bus stop, resulting in the failure of the bus to stop. In the latter case, this is because the process of buses stopping to pick up passengers is based on default assumptions of looking for – and being able to see – passengers hailing the bus from a standing position. Those who are seated in wheelchairs run greater risk of not being perceived as hailing the bus. The problems with the taxi and the bus travel occur quite frequently. In each case, two aspects of the experience are worthy of note in relation to mobile citizenship: waiting and having to find a plan B.

Bissell's (2007) research has focused on the practice of waiting as a means of positing a range of experiences beyond those suggested through the binary opposition of mobility/immobility. Waiting can be understood as a discrete kind of suspension. In some cases it need not be considered as predominantly a frustrated aspect of temporality but rather as a quiescent and discrete aspect of experience entailing disengagement, where people fill in time, often with the assistance of mobile devices. However, in Louis's case, in the context of his daily apprehension concerning the smoothness of the journey to work, there is no doubt that for him waiting entails a suspension of corporeal mobility which, if its duration extends beyond the amount of time envisaged, has to be actively endured and overcome.

> On more than one occasion I have wheeled off to the bus stop at 7:00 a.m. to catch the 301. I sit patiently with my headset on and on the arrival of the bus to find it not wheelchair accessible, as I watch all the other commuters board to head off to their jobs and school, I'm left waiting and hoping the next 301 is accessible.

Louis's account describes the experience of waiting beyond the expected time of catching the bus as uncertainty and "hope." The affective relationship to the journey that results from a set of technical and corporeal relationships means that uncertainty and risk may eclipse the choices entailed in acquiescence. In this context, there is always the need for Louis to have a plan B. This involves wheeling to the train station "in disappointment," and where getting the train involves wheeling 3 kilometers to work once he disembarks.

The case study so far has identified the number and complexity of space–time–technology relationships that need to align to allow for a relatively trouble-free journey to work. In spite of his planning and organizing, Louis knows that there is more that could – and has – gone wrong:

- The power supply needs to be guaranteed to enable charging of mobile devices and wheelchair.
- The support worker needs to be able to get to his house at 6.00 a.m. every morning.
- If a taxi has been ordered, an appropriate wheelchair-accessible taxi needs to arrive on time.
- If he plans to get the bus, the bus needs to be wheelchair accessible.
- Even if the bus is wheelchair accessible, Louis has to be seen at the stop so the bus will stop for him.
- If he cannot get the bus as planned, he must have a plan B – getting the train and wheeling 3 kilometers to work.

These aspects of the journey to work can legitimately be understood as decisive for a "productive" journey to work and already can constitute resources for the development of scenarios that at the very least might assist in mitigating uncertainty and risk. However, there is another key consideration here that technical supports also need to encompass: the symbolic experience of mobile citizenship.

Implications for mobile citizenship

Our relationship to mobility is central to the way we are now positioned with respect to mobile citizenship (Cass et al., 2005) and is a key stake in our access to contemporary cities. When our mobility is threatened, so is an aspect of our citizenship: our roles and social interactions are placed in jeopardy, and our capacity to maintain an appropriate "self" is destabilised. Thus, in addition to the points in the preparation for – and travel to – work where connections could be made more seamless and less fraught with uncertainty and frustration, we need to consider the role of the spatial–temporal–technical environment in the maintenance of "face": the positive social value a person can claim for themselves (Goffman, 1967).

For Goffman, people tend "to experience an immediate social response to the face which a contact with others allows" them, entailing an emotional investment in the presentation of face where their "feelings become attached to it" (Goffman, 1967: 6). Contemporary studies of mobility and the city attest to the ongoing relevance of such concepts, where the exercise of minor courtesies that maintain the face of others – such as respect, discretion, and sensitivity across our interactions and meetings – are the microcosms in which sociality is experienced and our sense of membership is sustained (Jensen, 2006). In this context, the capacity to sustain normality in everyday journeys such as getting to work is tied up with strategies and processes for reducing contingency and arbitrariness, which is, in turn, crucial for the capacity to control the face we present to the world (Misztal, 2001). For people with disabilities, the requirement to work around an environment that cannot be relied upon means there is no "time out" from planning. However, as we have seen, even excellent planning means that there are still no guarantees. Thus, in understanding Louis's journey to work we can also come to identify the way spatial, technical, and temporal aspects of everyday settings and sequences play a crucial role in mastering – or succumbing to – contingency and arbitrariness.

Louis's favourite place is the threshold of his house at the end of the day. It represents crossing into a world where much greater control, predictability, and comfort can be exercised. He provided an image of his threshold, followed by a caption (figure 12.1 below) that underlines the relief experienced in an environment where he is comfortable and has control over his environment.

The "entry to home," in particular, signified the predictable order of Louis's home, the knowing that in this environment he is safe and

Figure 12.1 My favourite place at my home is the entry. It's the entry to my space, my comfort. It's the entry to my stability. I know in my home I can live independently and free of too many fears. It's the entry to a place I can safely use my bathroom whenever I need to. I can sleep on my bed knowing I can safely transfer to and from my chair whenever I feel like a snooze. It's the entry to a place I know can make a coffee or a meal for myself. It's the entry to a place I most feel independent.

independent. Trust is established and maintained. The description also revealed that uncertainty and difficulty of urban life is on the outside of the threshold of the home. The home environment enables him to perform daily activities independently and autonomously as the environment is safe, supportive, and familiar.

Conclusion: Scenarios for more seamless journeys

This account of the experiences of journey to work has implications for the conceptualisation of transport systems from the vantage point of users' key expectations. These were described by Walker (2012) in terms of their capacity to address where and when people want to go, their value in terms of time and money, as well as safety, comfort, amenity, trust, and freedom to change plans. Understanding Louis's experience in the context of the time–space–technology relationships entailed in everyday journeys to work exposes the specific challenges faced by him when compared to people who do not live with disability. In particular, the case study illustrates the underlying ableism on which the transport system operates. As outlined by Chouinard (1997: 380), ableism refers to "ideas, practices, institutions, and social relations that presume able-bodiness, and by doing so construct persons with disabilities as marginalized...and largely invisible 'others.'" The example of accessing a public bus in this case study is a case in point, and explicitly reveals the notion of otherness at play, by requiring people with mobility impairments to have to preorder an accessible bus on their route 24 hours in advance, in addition to all the other additional requirements for organizing their daily journeys. Furthermore, Louis's experience also exposes the specific relationships entailed in the capacity to trust the role of transport in supporting the management of mobility (Audirac, 2008), even when he attends to the additional daily requirements for getting to work. As Walker points out, reliability of the transport system is crucial in this management. We have seen in Louis's experience the extent to which reliability is due, not only to features of the transit system, but to additional labour and effort in trying to mitigate the unpredictability and contingency of the journey to work.

It is the role of design scenarios to elicit insights from research and design practice and project possible or ideal contexts in which design might play a role. For the purposes of this chapter, they thus have a place in "prefiguring" (Willis, 2005) the various roles design might play in a more seamless journey to work on the part of people with disabilities. Here, it is important to address the broader questions of how design might serve the management of mobility on the part of people with disabilities. Further, however, it is also important that scenarios arise out of concrete stories of use (Willis, 2005) illustrated in the case of Louis's journey. This case example described routines and activities, mobility sequences, times and rhythms, and contexts: of physical–virtual connections or disconnections. Further, it illustrated the crucial symbolic importance of sustaining control over all of these routines, sequences, and connections as an important condition for mobile citizenship. This section identifies points in the two routines describe in the case study – "getting ready" and "traveling to work" – where it is possible to imagine newly configured relationships between people, objects, and actual

Table 12.1 Preparing to go to work.

Night before	**Day**				
various up to 9.30pm	**5.30 -5.50 am**	**6.00 am**	**6.05 am**	**6.30 am**	**7am - 7.45 am**
-Plug in phone and wheelchair, -Set alarms, -Organize clothing, -Pre-book taxis, or organise pick-up, or order wheelchair bus	Awake	Personal support worker arrives	Louis attends to personal care and dressing. Support worker attends to household tasks and clothes preparation.	Eat breakfast, might prepare snack /lunch alone or with assist from support worker	Leaving home to work. Time variation depending on type of transport.

and virtual environments that could enhance the experience of mobile citizenship for Louis.

Table 12.1 represents the key routines associated with getting ready that highlight the planning, time management, and technology relationships entailed in trying to achieve a "reliable" journey.

There are two sets of agencies involved in this: the activities associated with tending to devices (phone and wheelchair), and using devices to prearrange travel the night before; and the collaboration with the support worker. Louis currently lives with the daily risks associated with power disruption, meaning that the wheelchair and phone will not be charged, wariness about the reliability of taxis and buses, and the absolute disruption that occurs if the support worker cannot attend. In terms of implications for the development of scenarios that support the management of mobility, the getting-ready phase point to the importance of considering the following:

- Sociotechnical systems through which power backups can be put in place
- Transport planning and transport information that can be relied upon
- Communication networks that enable a support worker to assist at short notice

These are not merely discrete contingencies for which to plan, but operate together as a framework that is decisive for the experience of normality and trust in one's environment that is central to mobile citizenship.

The descriptions of travelling to work further highlight the level of uncertainty and contingency that attends the journey to work. This is illustrated below in relation to a possible occurrence in catching the bus – the arrival of a nonaccessible bus, and the requirement to resort to a plan B.

The differences between what is the expected pattern of a commute and Louis's experience outlines the scenario of current disrupted situations, where elements of access, predictability, time, speed, and social references, are "disconnected." In particular, the analysis of the conditions that produce the "bus not accessible" experience reveals that they pertain to documented universal design (UD) principles in relation to the physical characteristics of the transport environment that include "UD low-floor bus with levelled curb and UD-based bus stops, shelters, stations, and parking" (Audirac, 2008: 11) which would include provisions to ensure wheelchair users are visible by drivers. However, relationships of reliability clearly extend well beyond these physical provisions to the requirements for technical and organizational features of

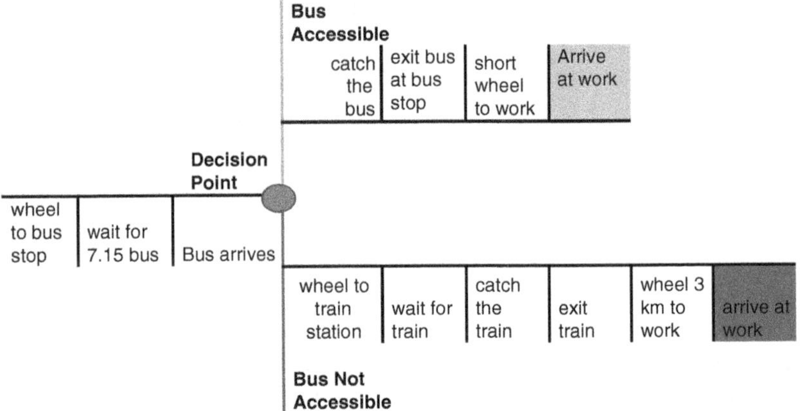

Figure 12.2 Diagram of connection and disconnection as a commuter for Bus 301.

journeys that are amenable to mobility management. Particular features of this management underlined by this chapter are the technical and organizational relationships that can

- assure users of the physical accessibility of the transport and the visibility of the passenger at the bus stop;
- provide a real-time system for identifying and accessing alternatives to an initially chosen transport option (requiring "Intelligent, real-time demand-response transit service with one call center" [Audirac, 2008: 11]).

Further, drawing upon these insights, design scenarios need to address the way these discrete aspects of the mobility system need to work together for individuals to enhance their capacity to trust their transport systems sufficiently to make the journey to work more sustainable.

The test of scenarios that take mobile citizenship seriously is the capacity turn experiences of "misfitting" into a more seamless sense of "fit." Rosemary Garland-Thompson's (2011) examination of these concepts in relation to in situ experiences of disability shows very effectively the way an experience of "misfitting" is so readily translated into "misfit" as a personal identity. Rather than opting for the projection of the relationship onto the bodies of people with disabilities, Garland-Thompson argues that experiences of misfitting are the very sites at which we should both problematize and enhance the way our environments operate as part of a democratic order. This chapter has argued that for scenarios to make a difference to designing for accessible cities, they need to inspire new design concepts that take seriously the capacity for normality and trust in our city environments as crucial for mobile citizenship. At the same time, they need to challenge the underlying ableism underpinning our systems.

Our chapter has illustrated the kinds of dimensions and their interrelations that need to be considered in promoting mobile citizenship. These include routines and activities, mobility sequences, times and rhythms, contexts of physical–virtual connections or disconnections, and the processes through which these are related to the symbolic experience of this citizenship. In this respect the scenarios should respond to the spaces, times, and technologies that are salient for participants in negotiating the city. We found that this is not limited to the journey to work, but entails other temporal frameworks and sequences as well: the notion of forward planning the night before the journey, the contingencies of power availability and ordering buses and taxis, and the longer duration of previous experience in getting to work which carries with it an ongoing sense of risk and foreboding about the things that may go wrong. Scenarios for more seamless journeys need to connect the dots of all these aspects of the experience.

Note

1 Muscular dystrophy refers to a group of conditions that cause wasting and weakness of the muscles.

References

Adam, B. (1995). *Timewatch*. Cambridge, UK: Polity.

Adam, B. (2006). "Time." *Theory, Culture & Society*, 23(2–3), pp. 119–138.

Audirac, I. (2008). "Accessing Transit as Universal Design." *Journal of Planning Literature*, 23(1), pp. 4–16. doi: 10.1177/0885412208318558

Bissell, D. (2007). "Animating suspension: Waiting for mobilities." *Mobilities*, 2(2), pp. 277–298.

Brighenti, A. B. (2012). "New media and urban motilities: A territoriologic point of view." *Urban Studies*, 49(2), pp. 399–414.

Carroll, J. M. (1997). "Scenario-based design." In M. Helandaer, T. K. Landauer, and P. Prabhu (Eds.), *Handbook of Human-Computer Interaction* (2nd ed.) (pp. 383–406). Amsterdam: Elsevier.

Cass, N., Shove, E., and Urry, J. (2005). "Social exclusion, mobility and access." *The Sociological Review*, 53(3), pp. 539–555.

Chamorro-Koc, M., Bucolo, S., and Adkins, B. (2012). "Using scenarios to explore the social aspects of design led innovations." In A. Villela (ed.), *Projecting Design 2012: Global Design Bridge* (pp. 186–191). Santiago, Chile: DUOC.

Chouinard, V. (1997). "Making space for disabling difference: Challenges ableist geographies." *Environment and Planning D: Society and Space*, 15, pp. 379–387.

Freund, P. (2001). "Bodies, disability and spaces: The social model and disabling spatial organizations." *Disability & Society*, 16(5), pp. 689–706.

Garland-Thompson, R. (2011). "Misfits: A feminist materialist disability concept." *Hypatia*, 26(3), pp. 591–609.

Gleeson, B. (1999). *Geographies of disability.* London: Routledge.

Goffman, E. (1967). *Interaction ritual: Essays on face-to face behavior.* New York: Pantheon Books.

Goffman, E. (1971). *Relations in public: Micro studies of the public order.* New York: Basic Books.

Hughes, B., Russell, R., and Paterson, K. (2005). "Nothing to be had 'off the peg': Consumption, identity and the immobilization of young disabled people." *Disability & Society*, 20(1), pp. 3–17.

Jensen, O. B. (2006). "'Facework,' flow and the city: Simmel, Goffman, and mobility in the contemporary city." *Mobilities*, 1(2), pp. 143–165.

Kitchin, R. (1998). "'Out of place,' knowing one's place: Space, power, and the exclusion of disabled people. *Disability & Society*, 13(3), pp. 343–356.

Litman, T. (2011). "Measuring transportation: Traffic, mobility, and accessibility." Victoria Transport Policy Institute, March 1, 2011.

Marston, G., and Lantz, S. (2012). "Policy, citizenship and governance: The case of disability and employment policy in Australia." *Disability and Society*, 27(6), pp. 853–867.

Misztal, B. (2001). "Normality and trust in Goffman's theory of the interaction order." *Sociological Theory*, 19(3), pp. 312–324.

Moser, I. (2006). "Disability and the promises of technology: Technology, subjectivity and embodiment within an order of the normal." *Information, Communication & Society*, 9(3), pp. 373–395.

Organisation for Economic Co-operation and Development (OECD). (2010). *Sickness, disability and work: Breaking the barriers.* Paris: OECD.

Urry, J. (2000). *Sociology beyond societies: Mobilities for the twenty-first century.* London: Routledge.

Urry, J. (2002). "Mobility and proximity." *Sociology*, 36(20), pp. 255–274.

Urry, J. (2004). "The system of automobility." *Theory, Culture & Society*, 21(4/5), pp. 25–39.

Urry, J. (2008). *Mobilities*. Malden, MA: Polity Press.

Urry, J. (2012). "Does mobility have a future?" In M. Grieco and J. Urry (eds.), *Mobilities: New Perspectives on Transport and Society.* Farnham, UK: Ashgate.

Walker, J. (2012). *Human transit: How clearer thinking about public transit can enrich our communities and our lives.* Washington, DC: Island Press.

Willis, A. M. (2005). "Scenarios, futures and design." *Design Philosophy Papers*, 3(1), pp. 1–37.

13

ExtraUrbia, or, the Reconfiguration of Spaces and Flows in a Time of Spatial-Financial Crisis

Bill Cope and Mary Kalantzis

The other than the urban

This chapter suggests that we might be entering a new phase of sociospatial development. In many respects, the shape of the emerging socioscape that we attempt to schematize in this chapter remains unclear. However, we want to focus on one possibility—that recent changes in the morphology of spaces and the dynamics of flows that in an earlier modernity had advantaged the "urban," may recently have begun to point in other directions, advantaging now what we propose to call the "extraurban."

We want to use the term "extraurbia" to highlight newly significant spaces geographically outside of the classically urban, spaces that have been comparatively neglected by scholars. We also use the term to indicate changed dynamics across these spaces, dynamics whose effects might be considered to be "urban plus"—most of what the heritage-urban offered to support human energies, and in addition new potentials for living and opportunities for social agency that traditional urban spaces today struggle to offer.

Each of the five spaces that we tentatively identify as exemplars of the extraurban are distinct from a sociospatial perspective. Each is peculiar to itself in a morphological sense. And each is at times bewilderingly variegated internally in social, cultural, and economic terms. However, we will argue that the characteristic "extraurbanity" these spaces share sets them apart from the structures and processes of the city of conventional understanding.

Central to our case is an analysis of the peculiar forces at work in the global economic crisis that began in 2008. This crisis was triggered by asset inflation which led to the overvaluation of primarily urban real estate. This structural and institutional catalyst has served to hasten the reconfiguration of spaces and directions of flows we map in this chapter.

This, by way of counterposition, is the underlying dynamic that characterizes the archetypical urban of our historical understanding:

Space 0: The urban

The advantages of the urban, at the most general level, have arisen historically from the pragmatics, sociability and aesthetics of collocation, contiguity, and propinquity. Grounded in the virtues and pragmatics of proximity, the urban has been the site of particularly intensive socioeconomic development (commerce, industry, employment) and cultural intensity ("civilizational" practices, cultural institutions, iconic edifices, and focal meaning-places). The engine of the urban, Edward Soja (2000: 12) calls "synekism" or "the economic and ecological interdependencies and the creative—as well as occasionally destructive—synergisms that arise from the purposeful clustering and collective cohabitation of people in space." Jane Jacobs (1969) speaks of the "spark of city economic life." Michael Storper (1997) explains the "vital role of agglomeration." A key motif in the long history of modernity, and longer still of "civilization," has been the drift from the country to the city. The end point of this tendency, Henri Lefebvre (1970/2003) predicted, would be a total urbanization in which even industrialized agriculture would sit in a subsidiary relationship to the city.

But what if structural and institutional conditions were to be reconfigured such that intensively physical-spatial agglomeration were to matter less? What happens when new aspatial modes of proximity are increasingly available and affordable, which could be put into motion for the purposes of production, community, and personal life? And what happens when some of the financial, environmental, and social costs related to intensive physical-spatial contiguity become prohibitive—a factor brought to our attention so sharply since the economic crisis which has come to a head since 2008? In these circumstances, we may be at an inflection point in the history of human settlement. New socio-spatial patterns are fulfilling many of the needs only served adequately in the past by traditional urban spaces. If this is the case, the model of extraurbia that we present here may challenge our conceptualizations of the urban/nonurban distinction and our understandings of centripetal forces and flows which historically favored urbanity.

This chapter explores the dynamics of the extraurban from two perspectives: the perspective of spaces—the five places we describe in the first half of the essay—and from the perspective of flows, or the synergistic interconnections between places that we analyze in its second half. Although our analysis finds continuities in the flows that stretch across the otherwise unfathomably variegated spaces of extraurbia, we don't mean to say that flows in any sense displace spaces as our primary unit of analysis and interpretation. For all the networkness and nodality of our contemporary global-local mesh, the flows we often call "globalization," we still mean to talk about spatially distinguishable places of being, of sensuous living and really imagined self-understanding.

The distinctiveness of the spaces we describe in this chapter are characterized not just by the forms of their connectedness with other spaces, but emerging forms of disconnectedness as well. For this reason, space matters again. The flows to which we refer take forms that are not just describable in empirical science, the processes of interconnection that are the Internet or commuting to work for instance, this tangible stuff of networks and their nodal routers. They take new, intangible forms of disconnection, of affirmative localness, of representational and cultural distinction. Flows give shape to spaces, they shape the dynamics of spatial becoming. While old connectednesses fray, new forms of relationality evolve. Place presents itself afresh as a site for defining these new flows. This is how we want to read a new dynamics of connection/interconnection that we hope to highlight through the urban/extraurban distinction.

Spaces of being: Five possible places of ExtraUrbia

Following are our five extraurban spaces, and after that, an analysis of the flows that define underlying continuities across extraurbia. At this tentative and conjectural stage in shaping the idea of extraurbia, we hesitate to present these as any more than propositions to be tested. The focal point of our empirical attention is the spaces and flows of the global north, and particularly those regions most affected, symptomatically, but the spatial diseconomies that triggered the global financial crisis. The epicenter of this crisis is, of course, the United States, which for this circumstantial reason alone is the primarily empirical base for the case we make here. By accident of history, the United States may well lead the world in the developments we describe. However, we contend that similar forces might be identifiable in a broad range of global spaces, even those less immediately affected by the current crisis.

Space 1: Edge-urban

The world's largest cities have become so large that at their edges, they are no longer viably urban in the traditional sense of integrated centralization and economies of physical contiguity. In the edge-urban, the centers of energy of people's lives become less tied by necessity to central cities. Instead, they are located in industrial "zones," office "parks," shopping malls, colleges, and recreational facilities dispersed throughout the edge-urban landscape (Dear 2000; Garreau 1991; Lang 2003; Lee 2007). A literature and mode of analysis has developed exploring their dynamics under the terminological flag "exurban" (Berube, Singer, Wilson, and Frey 2006; Spectorsky 1955; Taylor 2009). The edge-urban phenomenon also emerges in smaller cities, in their increasingly self-sufficient hinterlands.

By 2000, edge-urban development occupied 15 times the area of higher density urban development in the United States (Brown, Johnson, Loveland, and Theobald 2005). Between 2000 and 2007, roughly 3 million Americans moved to metropolitan edge-urban spaces (Kotkin 2009). Jobs, recreational opportunities, and cultural amenities are also increasingly locating in edge-urban spaces (Sööt, Berman, and DiJohn 2006). In this context, a trip to the city center becomes an occasional event rather than a frequent necessity.

Looking out from the city, edge-urban spaces—"exopolis" in Soja's terms—may appear to be distressingly fragmented places, sites of anti-urban "dispersed nucleation" (Soja 2000: 241). These places are often disparagingly called "sprawl" for the absence of the rigors of urban planning (Bruegmann 2005: 18). They may be places of refuge for people adversely affected by the spatial diseconomies that are the principal structural ground of the economic crisis. However, these are also places that can potentially deliver on material and cultural aspirations not so readily available in cities, providing at least interim respite from the sociospatial crisis of the city and improved living opportunities.

The outer limits of metropolitan areas are no longer bedroom communities for affluent commuters, the suburbanism of earlier times. Edge-urban spaces are home to a range of social classes and are increasingly self-sufficient in terms of providing a variety of employment opportunities and the necessities of everyday life (Kotkin 2009). In comparison to central cities, the edge-urban often offers better habitations for the poor, larger houses and gardens for the middle classes, and quasi-country estates for the affluent for the price of an apartment in the city. In other words, in a time of crisis, these may become places of relative hope and opportunity.

Certain types of edge-urban development have faced intense criticism, focused on concerns such as the fragmentation of habitat, inefficient resource use, destruction of prime farmland and ecosystems, and prevalence of negative social and health issues (Ewing 2008). However, an alternative perspective suggests that the costs of low-density development on the urban fringe have been overstated and their benefits, such as enhanced mobility, privacy, and choice for a larger segment of the population, have been granted insufficient recognition (Bruegmann 2005).

In addition to analyses that reveal a less destructive view of edge-urban landscapes, a shift toward environmentally focused values among edge-urban residents has been documented (Cadieux and Hurley 2009; Walker and Fortmann 2003). We also speculate that edge-urban and other extraurban spaces may provide improved opportunities for meaningful social interaction in comparison to their urban counterparts. While views from the outside perceive edge-urban spaces as socially isolating, ethnographic work has revealed a great deal of sociality in edge-urban developments, including forms of social unification and activism around mutual concerns, particularly focused upon environmental issues (Larsen, Sorenson, McDermott, Long, and Post 2007). Rather than breeding social isolation and wasteful consumption of land and resources as the predominant stereotype suggests, the edge-urban may foster relationships and perhaps breed social movements, including those that serve to enhance and protect the environment.

Slums or former slums might be considered another instance of the edge-urban phenomenon, places that have become mini-cities unto themselves, from the rudimentary planning of the "townships" of South Africa (Hart 2002), to the energetic village-like qualities of the slums of Mumbai or Rio de Janeiro, places of architecture without architects, of intensely extraurban human activity without social engineering or urban planning (Brand 2010; Davis 2006).

Space 2: De-urban

The de-urban consists of formerly urban, seemingly dead spaces in cities that appear to have been "hollowed out" and stripped of urban vitality. In actuality, they may in this moment increasingly be emerging as sites of potential and innovation (Davis 2002; Short, Hanlon, and Vicino 2007).

These spaces include collapsing neighborhoods in big cities—literally so, when abandoned and disintegrating buildings are demolished. At times, all that is left in these de-urban spaces is a checkerboard of buildings interspersed with "urban prairies." Or these might be smaller cities and towns that have imploded as their main industries leave. De-urban spaces appear at first glance to be places drenched in morbidly nostalgic regret.

Such a view, however, may be more from a perspective that romanticizes their intensively urban past than from that of their extraurban reconstructive potentials. These are spaces where, in an uncity-like way, houses, shops, and factories can be bought for well below their replacement costs, or rented very cheaply. Among the signs of new life in these places, we find welfare recipients who may have experienced displacement from gentrifying urban neighborhoods, working-class families in search of more spacious housing, arts and craft colonies emerging where there is next-to-no home or studio overhead, middle-class people renovating decayed mansions, neighborhoods being reclaimed, and abandoned commercial strips filled out with secondhand stores, cafés, and galleries.

While gentrification and ultimately social homogenization due to the displacement of those unable to afford rising rents or property taxes is a possible path for the de-urban, the potential for something different is also great. In his 2010 documentary, *Requiem for Detroit?*, Julien Temple presents a view of utter abandonment and ruin in Detroit while highlighting signs of rebirth. Temple shows an innovative and hopeful population making use of abandoned spaces, reoccupying buildings transforming them into farms sites left vacant by demolitions. An urban agriculture movement is burgeoning and de-urban spaces throughout the United States (Lawson 2005). These are early signs that in response to their own and the broader sociospatial crisis, de-urban spaces may give rise to new types of economies and new ways of living together (Schilling and Logan 2008).

As de-urban spaces become extraurban, they may regain population. However, in most cases they are not likely to return to the population they had in their urban pasts. Therefore, de-urban spaces tend to be much less dense than their traditional urban counterparts. This can have several effects. The de-urban can serve ecological functions that are not feasible of the classically urban. For example, when previously urban spaces are abandoned, the reduction of impervious surfaces allows for more effective storm water infiltration. Abandoned spaces allow plant life to flourish, which can in turn provide new wildlife habitat. An estimated one third of Detroit has been reclaimed by plant life (*Requiem for Detroit?* 2010).

While much of the extraurbanization of the de-urban has been bottom up, top-down action is also playing a role in some places. "Right-sizing" initiatives are being established in so-called shrinking cities, which by definition have lost at least 25 percent of their population (Schilling and Logan 2008). These programs and policies focus on reimagining rather than restoring places that have experienced great population loss and disinvestment. They may promote extraurbanization of de-urban spaces (as opposed to continued disinvestment or gentrification) by implementing more efficient and cost-effective infrastructure and municipal services and reusing vacant properties in ways that offer new social opportunities and community-based economic development.

Space 3: Micro-urban

Towns and cities of 10,000, 50,000, or even 200,000 might with some justification be considered not to be archetypically urban. Or at least they do not meaningfully sit on the same scale as large twenty-first century cities. Yet micro-urban spaces are sites of dynamic growth. They have been less adversely affected by the spatio-economic crisis that began at the end of the 2000s. The overall trend in domestic migration in the United States has recently made a dramatic shift toward small towns. Growth in nonmetropolitan areas now exceeds metropolitan growth in the United States (Bell and Jayne 2009; Cox 2008).

Unlike "small towns" of the stereotypical imagination, micro-urban places have increasingly fluid and diversified populations, both in terms of socioeconomic differentiation and the ethnic origins of newcomers (Miraftab and Mcconnell 2008). Black–white segregation declined in U.S. towns in the 1990s more than it did in cities, and segregation patterns are typically less distinct in newer edge-urban and greenfield developments (Lichter, Parisi, Grice, and Taquino 2007). Diversified employment opportunities are locating in micro-urban locations where property and labor are significantly cheaper than large cities and a better "quality of life" may be attained (Johnson and Rasker 1995). The types of employment that are growing in micro-urban areas include knowledge and service industries as well as manufacturing (Quark 2007; Vias and Nelson 2006).

Apparent similarities between large urban centers and micro-urban spaces do not necessarily mean that outcomes will be identical. Many small towns are successfully working against problems like unsustainable land use and socially and economically divided populations (Mapes 2009).

Space 4: Greenfield

Beyond the edge-urban and outside of the micro-urban are various forms of "greenfield" life, in rural hamlets, on farms, in holiday houses, in retirement villages, in forest cabins or beach shacks, and in caravan and mobile home parks. Growth in rural counties in the United States has increased dramatically in recent decades, particularly in high-amenity areas (Cadieux and Hurley 2009; Gosnell and Abrams 2009; Gunderson, Pinto, and Williams 2008; Johnson and Cromartie 2006; Krannich, Petrzelka, and Brehm 2006; Rasker, Gude, Gude, and van den Noort 2009). New patterns of employment and socioeconomic diversity are having a large impact in rural areas (Woods 2009). In addition to amenity migrants, lower wage workers are moving to greenfields to work in service jobs and in factories. Industry is purposefully locating in extraurban greenfields, archetypically in contrast with its formerly urban "rust belt" locations (Phelps, Wood, and Valler 2010). In many parts of the United States, Latinos are taking these jobs. The Latino population has increased dramatically in rural areas, comprising almost half of the total growth between 2000 and 2006 (Parsi and Lichter 2007).

Located within what is typically associated primarily with intensive monoculture farming and a conservative rural lifestyle, greenfield spaces have also come to be associated with images of nature, peace and quiet, space, family friendliness, and community belonging, often due to marketing campaigns to attract tourists and new residents (Berry 1976; Champion 1992; Gkartzios and Scott 2010). In-migrants from cities overlay characteristically urban expectations such as gender equality, social mobility, cosmopolitanism, and environmental concerns (Jones, Fly, Talley, and Cordell 2003; Munkejord 2006; Qviström 2007).

Greenfields, along with other extraurban spaces, are increasingly multicultural (Agyeman and Neal ; Miraftab and Mcconnell 2008; Singer 2004). While residential segregation patterns in rural areas and small towns may at first glance appear similar to those found in large cities, extraurban greenfields can also provide advantages for immigrants over traditional gateway cities and even a loosening of the rigidities of spatial segregation (Lichter, Parisi, Grice, and Taquino 2007). Greater economic integration is one advantage for immigrants that has been documented in small towns (Bernard 2008). Evidence also points toward improved social integration in extraurban greenfields: in one study, adolescent residents of two rural towns in the United States have been shown to exhibit positive attitudes toward Latino immigrants in their communities, much above what was expected (Gimpel and Lay 2008).

Space 5: Off-the-grid

In formerly remote places—in mountains, forests, coastlines, and deserts—off-the-grid energy sources and online and physical deliveries make it possible to live virtually urban, socially and culturally proximate lives (Ryker 2007). These are also spaces for increasingly autonomous yet globally integrated indigenous or First Nation communities. In the extraurban dynamic, "the remote" becomes a relational concept, increasingly connected in its difference to the wider world while at the same time being defined by its radical disconnectedness (Clancey 2004).

It has become a canon of conventional wisdom that people are moving into cities. However, the data on which these assumptions are based may in fact aggregate in such a way as to obscure other trends. For instance, such demographic data as we have show an overall, albeit uneven, increase in migration to these types of spaces and a concurrent decrease in migration to large cities (Champion, Hugo, and Champion 2003).

We would like to suggest this possibility: if we were to refigure the demographics of urbanity, if we make the distinction between urban and extraurban spaces that we are now suggesting, we may well discover a trend that points toward an imminent reversal of the long-standing drift of the world's population to cities. The data may not quite yet be figured in such a way to sustain such a claim. At the very least, we want to propose a research agenda that refigures the data of human spatial distribution.

Flows of becoming: Continuities of connection and disconnection across the places of ExtraUrbia

What characteristics might these five spaces share, notwithstanding their extraordinary variety and the fact that they are themselves deeply differentiated internally? We turn now to analyze flows, and particularly those transformational dynamics of connection and disconnection that distinguish the newly extraurban from the anachronistically urban. We classify these into three broad categories: ontological flows, flows of conviviality, and representational flows.

1. Ontological flows

Ontological flows are the stuff of materiality, of buildings and food. For all the talk of information society, knowledge economy, and the postmodern ascension of the sign, enduringly we need to eat, we dwell in housing, we use and often also like sensuous things. Things have to get to us, and this getting is a viscerally ontological thing. After the excesses of postmodernism and the hyperbole of postindustrial development ideologies, we want to reground our analysis in these so-ordinary things.

Flow 1.1: Propertyscapes

To start with one of the banalities of dwelling: real estate prices. These have been a trigger for a cascade of profound consequences, pivot point in the economic crisis that began at the end of the 2000s. At the simplest level of analysis, we need to consider the cost of space and in particular the differentials between the cost of urban and extraurban space. Here, our focus is on two kinds of flows: flows of people from urban to extraurban places of dwelling; and flows as magnetic/repulsive systemic forces of connection/disconnection, in this case the radically differentiating real property economies of the urban vis-à-vis the extraurban.

Real property prices in the places of extraurbia have over the past few decades become relatively much cheaper than urban property. Frequently, in fact, real estate prices are at or below construction costs, and particularly so since the coming to head of the property-financial crisis of the late 2000s (Glaeser and Gyourko 2002; Krugman 2005). This is a global phenomenon, and twofold development. On the one hand, starting in the last decades of the twentieth century, urban property prices have become exorbitantly high, and even when extraurban prices have risen, it has mostly been at a slower pace. The urban/extraurban differentials have become even greater since the generalized real estate price deflation at the core of the crisis. And while the urban-extraurban cost gap has grown, before and after the crisis, the historic advantages of urban physical-spatial proximity have waned. As a consequence, the costs of space in cities are no longer such a matter of necessity for households and employers.

To step back in order to interpret these changes in a wider frame of analysis, monetarist economics, obsessively interested in inflation, selectively removes from its calculations the largest item of household consumption—the capital price of housing. There is no practical, material reason why it should as many consumer durables have the same qualities as capital—they last for years, they depreciate over these years, and their costs should thus be amortized. These similarities are conveniently ignored. So is the fact that property depreciates, not only in terms of the half life of fixtures and decoration, but the evaporation of monetized value in times of bust and also in larger structural frames of reference when population declines in neighborhoods, a process reaching its extremes in the case of the near-valueless properties of de-urbia.

The reasons for leaving real property out of inflation counts are ideological. They support a system-defining delusion. To include property would convey bad news that neither homeowners nor the grandees of financial capital want to hear—of dangerous inflation in one moment of the boom–bust cycle and dangerous deflation in the subsequent period of stalled recovery. Not only did equity gambling on inflation of real property prices become central to the psyche of a class of homeowners greatly enlarged by the boom, but it also became a key to the derivative forms of gambling that are the basis of the whole superstructure of securitization that is the contemporary financial system.

During the bubble economy of the post–Cold War decades, and without factoring in housing, the monetarists allowed their false assumptions to be a cause for self-congratulation, their achievement of having engineered low inflation. In the case of urban spaces, the imaginations of homeowners and bankers alike came to be acutely separated from ontological realities—the land value of alternative sites plus the costs of construction. Hence, we witnessed a phantasmagoria of speculation and secondary consumption securitized against illusory gains in appraised values.

The lie about low inflation in the boom meant that the monetarist economists could reassure employers of the success of their policies because businesses enjoyed a low rate of wage growth which, as the theory goes, is appropriate to low levels of inflation. However, this created a radical disjunction between the inflated cost of housing and homeowners' capacity to pay.

Bringing the purchase cost of housing back into the equation, the reality is that urban housing has been a site of system-destabilizing inflation, a process that in fact lasted for some these decades. By the early 2000s, urban housing practically became unaffordable for the poor and middle classes alike. The negative effects of declining affordability were counterbalanced by a global financial industry willing to fund deficit lifestyles against the appraised value of housing and in which housing had become a principal cause of the household deficit.

Then the obvious—but unthinkable for most financiers and homeowners alike—happened. The urban real estate bubble burst when debt service gearing reached a breaking point for a large number of borrowers (Quiggin 2010). The result was subsequently dramatic price deflation, again masked by the noninclusion of the capital costs of housing as an item in consumer price indices.

The fall in urban and extraurban property values has been uneven. However, as a general rule, while many city prices have remained unrealistically high, prices in many extraurban spaces have remained relatively more affordable and with price deflation there as well, even more affordable. Because access to land is less constricted and homebuilders face relatively fewer planning restrictions in extraurbia, housing prices there retain a more reasonable relationship to the cost of manufacture of housing.

When governments in several countries reacted with bailouts to prevent the collapse of finance capital, it was not just that the banks that were too big to fail, but the housing-inflation economy on which they had increasingly built their businesses. For the moment, the banks have been propped up; however, the crisis only served to accelerate the shift from the urban to the extraurban that we describe in this chapter.

The bank bailouts may have temporarily arrested the slide, funding the losses of the banks incurred by the capital deflation that they failed to anticipate as a sequel to ungrounded capital inflation. However, this is only a temporary palliative. Housing in large cities still does not bear a workable relationship to household incomes. For all the talk of a putative "knowledge economy" (Peters, Marginson, and Murphy 2008) where the real value of productive enterprise is in goodwill, brand value, human capital, and the like, the financial system's primary form of securitization was—and remains to be—real property, mostly domestic housing, and mostly grounded in values dangerously disconnected from incomes.

The worst and most system-destabilizing disjunctions remain cities, where, thanks to the bailouts, prices have only fallen from hopelessly unrealistic to plainly unrealistic. The unavoidable response is an acceleration of people flows to extraurbia. If, as an overcharged and debt-stretched housing consumer, you want an alternative to the still-too-high urban values, then make the move to extraurbia. This is a simple and practical matter of "affordability" common to extraurban spaces.

Flow 1.2: Pathways

Here are some characteristically urban flowpaths:

1 Commute by train or car to work, suburb to the city, from one spatially specialized locale to another
2 Drive to downtown or to a large mall to shop
3 Drive or catch a train to the city to go to the cinema or wander a market or visit a big-city bookstore

Here are their extraurban analogs:

1 Telecommute because you are an online teacher or because you are a designer who works from a home office (Mitchell 1999); or travel

less because your person-to-person work does not require you to travel to work every day; or travel a short distance because homes and workspaces are collocated in mixed developments, or closer differential zoning. While only 3.8 percent of the U.S. population worked from home in 2000, this is a considerable increase from previous years and the trend is growing (Pisarski 2006).

2 To shop online is to turn a privatized flow (drive to the shops) to a socialized flow using shared communications and transportation infrastructure, and that as such can be much more efficient in terms of time and energy use. Sometimes the product is a frictionless download away, through shared infrastructure of the social web (a song, an e-book, a movie). Given the remarkable efficiencies of this new mode of production and distribution of cultural contents, this is often surprisingly cheap, even free. Other times, physical delivery is through the burgeoning public transport delivery system, the inexpensive and efficient shared transport system of trains, planes, and delivery vans, a system that with computerization has continued to make advancements in efficiency and affordability (Glaeser and Kohlhase 2004). E-commerce in the United States grew 11 percent in 2009, despite recession and stagnant retail figures, to reach 7 percent of all retail (Fowler 2010).

3 Or, go to the nearby small-stall market, and you'll not have to go far, because markets making virtue of their convivial localness and low costs are proliferating. All of these developments change the dynamics of spatial relations.

Despite its spatial dispersal, extraurbia may well be a geography of driving less. It may represent new transportation efficiencies. In-person travel—to a meeting (when not for a virtual meeting), to an in-person class (when not in an online class), to an aesthetically different recreational space (when not on documentary TV or video)—can then become a matter of now-and-then choice rather than daily necessity. When one does need to travel, distance is less a detractor from the other benefits of extraurbia when access to well-serviced regional airports is more readily available and relatively inexpensive (Rasker, Gude, Gude, and van den Noort 2009).

Perhaps, too, a new phase of transportation technology development might extend this process of socialization: the GPS route planners which steer drivers away from urban congestion; the impact detectors and road-based guidance devices that may soon regulate traffic as well as reduce traffic accidents (Mitchell, Borroni-Bird, and Burns 2010); even one day perhaps, the kind of hybrid public–private personal rapid system foreshadowed by the failed Aramis project of 1970s and 1980s Paris (Latour 1996). In all these scenarios, the logistical practicalities and efficiencies of contiguity that characterized the city may prove anachronistic. So, paradoxically, a characteristic feature of the shift to extraurbia may not be more transportation across greater distances and private transportation, but reduced physical movement of people and goods, and increasingly coordinated or socialized systems of transportation.

Flow 1.3: Extrastructures

The infrastructures of the classically urban were based on large energy distribution grids, remote waste treatment facilities, and distant water sources. The dominance of these infrastructures, which are grounded in the logic of economies of large scale, may come under challenge from a reversal if infrastructural logics that we will call "extrastructure." These include decentralized, relatively autonomous sites of energy production such as onsite solar, wind, or geothermal energy on or off the grid, efficient gray water recycling, rainwater collection, and rubbish composting. Alternative, localized modes of energy and water provision are rapidly becoming favored over traditional infrastructures in areas where no infrastructure previously existed (Ness 2007).

As these new technologies develop and become cheaper, they could present a practical and more affordable alternative to grid infrastructures, even in places where access to traditional infrastructures currently exists. Extrastructures may be particularly attractive replacements for aging infrastructures that are increasingly expensive to maintain in deurban places that have experienced a decline in their tax base through population loss (Schilling and Logan 2008).

As the social will to protect the environment continues to gain momentum, popular interest in extrastructures may increase as well. The renewable energy standard included in the 2009 American Clean Energy and Securities Act and some state renewable energy mandates may provide additional incentives for technological advances in alternative energy.

Flow 1.4: Productionscapes

In recent decades, production has shifted away from a centralized urban model and become relatively more dispersed. Nonmetropolitan employment has diversified greatly since the 1970s (Vias and Nelson 2006). Extraurban spaces are quintessential sites of new and old production—from the innovation industries of R & D and design, to new information sweatshops such as call centers, to high-tech manufacturing, to relocated old-economy industries such as handcrafted furniture and abattoirs. Nonmetropolitan areas are particularly attractive for food manufacturing (Schluter and Lee 2002). Wages tend to be lower in extraurbia than in urban centers (Johnson and Rasker 1995), but the costs of living (primarily housing) are lower, too. If the lower costs are lower than the lower incomes, and this is often the case, this results in a better standard of living at all levels of the labor market.

From the point of view of the enterprise, extraurban locations are "competitive" for their affordable properties and the flow-on of reduced housing costs into labor markets. This is the case for microbusinesses, midsize, and very large enterprises. In fact, this dispersal of production may need to be added to the feature set of what Pieterse (2008) calls the "new globalization." From the point of view of employees, if wages are somewhat lower, this is more than compensated by the availability of employment and the cost of living. Paradoxically, senses of locality and community may also provide leverage for organized labor (Nelson and Hiemstra 2008), even across historic ethnic and racial divides. And from the point of view of nonworkers, people on pensions or welfare, these are relatively inexpensive places to live, and more pleasant for the price. So, across the various spaces of extraurbia we may see the development of new modes of production, at least subtly different in some significant ways to the modes of production characteristic of cities.

Flow 1.5: Consumptionscapes

New modes of production spawn new modes of consumption. The focal point of consumer energy during the twentieth century shifted from one iconic urban site to another, from the downtown with its main street to the shopping mall and the big-box stores at the edge of the inner city or in the middle of the suburbs. Inefficiencies and cost structures may, however, may spell the end of both malls and big-box stores (http://www.deadmalls.com/).

The Internet is taking a newly important place in the consumption practices of more people (Fowler 2010). Online shopping works particularly well in extraurbia, providing significantly more consumption options than those available in bricks-and-mortar stores in big cities. Access to the Internet remains uneven, although, according to a recent PEW Internet Life survey, the digital divide is closing (Jones and Fox 2009). As more people gain access and more information and services are provided online, those who do not have access are at an even greater disadvantage (Tongia and Wilson 2007). However, accessibility is hastened even in the poorest neighborhoods and parts of the world by cheap computers and mobile devices that can access the Internet through light-weight Wi-Fi infrastructures.

New modes of online consumption emerge that provide the inhabitants of extraurbia not just the same depth of consumption possibilities as people in big cities, but considerably more than that. Megastores like Amazon have a breadth of inventory with which no bookstore or department store from the era of spatially massed shopping could ever compete. Extraurban consumptionscapes also include the myriad of specialist stores with narrower focal points than retailers of the recent past. These stores have a depth of product range, online information, and live help that no conventional specialist store could ever offer. They run on databases with filter mechanisms which means you can sensibly sort what you want from 100 dimmers (http://www.dimmers.net/), 1,000 light bulbs (http://www.1000bulbs.com/), or 5,000 faucets (http://www.faucetdirect.com/). This brings products to light that no browsing along physical shelves ever could.

And here is yet another retail model in these new consumer economy: the micro-manufacturers who produce on demand—the small metal shop manufacturer making stainless steel sinks who delivers them to order for much less than the big manufacturers, the cutting board manufacturer who will make boards to any size you order, the artists and craftspeople with online galleries, the boutique vineyards who sell their wine online and globally. These enterprises represent a radically new structural model, the ultimate form of disintermediation in which the finely niched manufacturer delivers direct to the distant consumer, with no distribution warehouses or retailers conducting costly middle-person services that require expensive supply infrastructures. In this way, myriad new enterprises reconfigure the supply chain in quite fundamental ways, cutting out many of its more expensive layers of warehousing, distribution, and physical retail display.

Things in this economy become cheaper for the poor and the more affluent alike. The "new economy" is not just the symbolic facade that are the ephemeral signs of a website. The key is the new, radically disintermediated material production infrastructure that the Internet makes possible. Then there is the e-Bay economy which blurs the very distinction of consumer and seller, and creates a market agnostic to retail scale, a place for miniscule sellers offering their products in the same space as e-commerce giants.

Importantly for the spatial arguments we are making in this chapter, retailers don't need to be near you—they can be, and are, located most competitively in the least expensive reaches of extraurbia. They offer a deeper, broader, more engaging consumer experience, and, for the money, they offer more of it. Of course, these new modes of consumption are available to city dwellers, too. However, at a particular level of income, people in extraurbia can do more consumption within the resources available to them, and city dwellers have lost their costly historic advantage of being close to "good stores" based on larger markets or "good-value stores" based on economies of large scale.

Flow 1.6: Socioscapes

In an earlier modernity, spatial separation created social divides and racial and ethnic separations. Spaces that have since become extraurban may have formerly been places of "white flight," of small-town insularity, of cultural and demographic stasis, or of rigid class segregation. They may have been places for those who could afford it to escape from the city, which seemed from an outsider's perspective to be swarming with immigrants, conflicted by the claims of social movements, afflicted by social problems, and made dangerous by extreme inequality and the agglomeration of poverty into urban ghettos.

If extraurbia offers advantages to everyone, patterns of social and cultural division may be changing. Extraurban spaces could become sites of opportunity and improved lifestyle for all—for refugees, documented and undocumented immigrants, the poor, the middle classes, and the affluent. For this reason, too, places that had been demographically homogenous are becoming cosmopolitan. These changes will not of course be without the pains of racist reaction, or domestic dislocation, or intergenerational disjunction, or social anomie—all of which, however, are likely to appear in different guises to those we have associated traditionally with city life.

The extent and nature of the human variegation across extraurban communities is, paradoxically, a continuity of flow that defines all their socioscapes. To make a bold suggestion, and notwithstanding the multiple points of crisis and distress, perhaps the transition to an extraurban social order could be faster and less troublesome than similar transitions in the history of social transformation that accompanied classical urbanization. While evidence of interaction and integration among different groups that come together in extraurbia is varied and is dependent on a variety of factors, there are signs that such a social transition is taking place. One immediate sign may be the emerging phenomenon across extraurban spaces of increasingly socially integrated communities.

Flow 1.7: Ecoscapes

From an environmental point of view, the extraurban can at times be a site of particular horrors. Out of sight is out of mind in cases of mountaintop mining, aggressive industrial farming practices, paving over prime farmland, destroying wildlife habitat, and large-scale burning of fossil fuels to generate electricity, for instance. However, extraurban spaces also offer peculiar opportunities for the reconfiguration of human–ecosystemic relations. The introduction of new environmental technologies is one such opportunity. Extraurbia is especially well suited for the construction of the "extrastructures" of post-grid, a-nodal energy production. On-site composting and gray water recycling are easier alternatives in extraurbia, and potentially cheaper per capita than the waste disposal infrastructures of big cities. Water can be collected on site. Extraurbia, in other words, can be a place where it is more practicable and cheaper to implement certain green technologies. In addition, the sites of primary engagement for human–ecosystem reconfiguration, and thus the burden of environmental responsibility, falls primarily in extraurbia, for this is where the energy for old cities is still generated, the food produced, the building materials sourced, the sewerage pumped, and the rubbish dumped.

From the perspective of the city, the environment is an external site of referred pain, a site of collateral damage. For this reason, it is a site of merely abstract concern. Extraurbia, by comparison, is often understood from within as dwelling that embodies nature. Quite simply, it as a place where people's environmental sensibilities are shaped by being closer to nature. In extraurbia, the questions of food production, energy generation, water qualities, and waste disposal are viscerally localized.

The obverse of this spatial peculiarity is the empirical character of the human flows from the urban to the extraurban. Amenity-influenced migrants tend to value preservation of nature that supports the quality of life that they obtained by moving to extraurbia. They have also been shown to be more likely than others to be associated with conservation groups. It has been suggested that they help strengthen these organizations by providing enthusiasm and increased income through membership dues, contributions, and fund-raising campaigns (Levitt 2002). In addition to amenity migrants, economic migrants and long-term residents also tend to have a sense of environmentalism, though this means different things to different groups. For example, conventional farmers, while often derided for their practices, often pride themselves for their profound connection to the land and its life cycles. They have a stake in protecting the land from which they make their livelihood (Sullivan, McCann, and De Young 1996). International migrants to the United States, who are increasingly moving to extraurbia, have also been found to demonstrate higher levels of environmental concern than native-born residents and to exhibit consumption practices that are protective of the environment (Hunter 2000; Pfeffer and Stycos 2002). For these reason, we speculate that extraurbia may become the focal site for activism and action in the creation of new ecoscapes (Larsen et al. 2007). Common to extraurban spaces, in other words, there may be both a heightened sense of present environmental crises, as well as congenial sites for the development of innovative, green practices.

2. Flows of conviviality

Flow 2.1: Governance

Extraurban spaces may be absorbing some of population that has been pushed out of city centers by gentrification and increasingly punitive neoliberal governance (Harvey 2005). They may also at times be sites of neoliberalization themselves. However, it is possible that extraurbia may in some respects be better equipped to respond to neoliberal challenges in more humane ways. For their dispersal, for their comparatively small scale, for their relative autonomy from heavy urban-centered governance structures, and for the relative informality of their institutions of civil society, the spaces of extraurbia may allow the possibility of more devolved, flexible, and responsive modes of governance. Smaller governance structures have been shown to be better able to efficiently address local needs, and with less financial expense (Cox 2005, 2008; O'Toole and Burdess 2005).

The neoliberal down-scaling of responsibility for personal well-being that has devastated urban residents who previously relied on social services may also be helping to bring about a reemergence of community in extraurban places (O'Toole and Burdess 2004). The neoliberalization of governance and the particular social mix and culture of extraurban spaces may come together to stimulate innovations in organization, collaboration, and mobilization that result in the emergence of a new kind of social capital that align specifically with extraurbia (Laliberte 2009; Nelson and Hiemstra 2008; Pink 2009). Compared to the city, the relative institutional thinness of extraurban spaces presents dangers—of hyperexploitation, neoliberal lawlessness, and poor planning. But on the other hand, extraurban spaces may also offer possibilities in the form of "quiet encroachment" of participatory self-governance and what might be called, for their unassuming activism, "social nonmovements" (Bayat 2002).

Flow 2.2: Identities

Historically, cities sorted demographics into spatially distinguishable neighborhoods, or quarters, or ghettoes. Extraurban spatial sorting may prove to be less spatially and socially rigorous. This is in part supported by a broader trend to viable postterritorial identities, ending the conventionally framed isomorphisms of space and social form and the ascription of cultural authenticity or essence to space (Massey 2005). In the era of digital communications and online community, person-to-person collocation is less needed in order to maintain diaspora, or knowledge community, or fashion, or fad, or fetish. Here, we see a dynamics of difference emerging that is less determined by space, and for this, it is more complex and multilayered (Kalantzis and Cope 2009).

On the other hand, identities continue to be tied to place. Having a home in the country, for example, or living "off the grid" are key to the identities of some. And as extraurban spaces compete for investment, visitors, and new residents, many of them play up certain aspects of their localness, which are often incorporated into the identities of the people who have links to those places.

Flow 2.3: Communities

Every demographic has its peculiar reasons to move to the spaces of extraurbia—retired people for a quasi-vacation lifestyle, families for their children, gays for places of shared style, immigrants and refugees for an entry point into the labor market. Increasingly, extraurbia becomes a place of cosmopolitan community. Their newly acute juxtaposition will inevitably produce dynamics of dissonance and new points of clash. However, for every moment of dissonance, we may witness the rise of newly constituted intersectional rainbow coalitions, as diversifying communities face cutting-edge workplace, environmental, or educational issues.

According to some, the sociality and collectivity that is traditionally a feature of the urban has been eroded by individualization (Harvey 1996). Perhaps some of what has been lost may be re-created in extraurbia. This is evidenced in part by the increase in demand for local food, locally owned and operated businesses, and localized experiences. While the rise in desirability of all things local could be deemed inauthentic authenticity, new points of closure by insisting on homogeneity as distinctiveness, this phenomenon could equally manifest itself as a pluralized, cosmopolitan localness. Whatever the tensions in "authenticity," the result may be a productive dissonance that generates increased social interaction and conviviality among socially diverse neighbors.

3. Representational flows

Flow 3.1: Communications

In the new communications environment, sharing of meaning becomes less dependent on the heritage synergies of collocation or economics of contiguity. Facebook creates a never-before envisaged shape of proximity in life narratives. Voice-over-Internet and videoconferencing remove diseconomies of distance. Mobile phones "roam" as if location were immaterial. People in cities have no better access to good newspapers than people living beyond their print distribution reach. In any event, mass market newspapers, grounded in economies of large scale, find they are competing with disruptive economies and qualities of small scale—the thematically particular blogs, the micromedia that cover a few hundred households, the slicing and dicing of information in blogs that reaggregate and link from one perspective or another. In all of these respects, the city loses its communicational advantages (Collins and Wellman 2010). In the words of Saskia Sassen (2006: 328), digitization is responsible in part for the "destabilizing of older formal hierarchies of scale and the emergence of not fully formalized new ones."

Flow 3.2: Innovation

Where does innovation increasingly occur? On college campuses that have for a long time been located outside of big cities, in towns that are distinctively attractive for precisely that reason. Or in the research parks that incubate enterprises spun off from university-originated IP. Or in the new economy multinationals that are headquartered outside of big cities or that have their R & D divisions located outside of cities. Or the R & D start-ups that take advantage of lower overheads and regional incentives commonly offered in one or other of the spaces of extraurbia.

Flow 3.3: Learning

And finally, how is knowledge to be transferred intergenerationally? The answer will in part be through environments of ubiquitous learning, ranging from online degrees, to small local schools relying on online infrastructure, and beyond the conventional classroom or training room, on and through networked mobile devices, where learning can happen any place and anytime, just enough and just in time (Cope and Kalantzis

2009). As the spatio-institutional walls of the traditional school come down, there need be no location-determined educational disadvantage. Online education has been shown to provide higher learning outcomes than traditional education (Bernard, Abrami, Lou, Borokhovsk, Wade, Wozney, Wallet, Fiset, and Huang 2004; Shachar and Neumann 2003). Twelve percent of all students enrolled in degree-granting university programs in 2009 were enrolled in fully online programs, a figure that is predicted to continue to rise (Shachar and Neumann 2010).

Toward an ExtraUrban normativity

The ontological, convivial, and representational flows we have sketched in this chapter are constitutive of continuities across extraurban spaces, distinguishing the newly extraurban from the historically urban. These flows reflect new and complex patterns of uneven development, patterns that are no longer reducible to conventional spatial mapping. They also represent sites of opportunity for persons who for varied reasons need or want to escape the urban epicenters of the current spatio-financial crisis. The flows we have described here are ontological, viscerally connected to our human natures as creatures who dwell, and the things we do today to shape the conditions of our dwelling as best we can under prevailing historical conditions. They are convivial, going to the heart of our interrelationships as sociable beings. And they are representational, the stuff of our epistemic selves, our meanings, our envisagings, and our imaginaries. The flows are both tangibly material and at the same immaterial force fields that have material effects. They include new processes of pointed disconnection as often as relations of connection. They are multiscalar, detectible in the microdynamics of intensely localized activities as well as within regional and global frames of reference. They are deeply structural at the same time as they involve interacting human agents—between the well-placed purveyors of institutional power, for instance, and those attempting to sort out their lives afresh under new conditions of spatiality.

These are disconcerting times. Who, until recently, would have been even able to imagine an "aspatial globalization" (Massey 2005: 81) or the "loss of diagrammatic clarity" (Bruegmann 2005: 49) manifest most acutely in these extraurban spaces? Who could have imagined that the city might cede many of its magnetic advantages to the not-city? However, we have attempted to argue in this essay that such a day may soon be arriving. And when it does, we might also be able to transfer lessons being learned in extraurban spaces to urban spaces, ideas that will make our cities better dwelling places. The peculiar flowscapes of the extraurban make it a vital source for social transformation in a time of spatio-financial crisis.

If our schematization of the dynamics of extraurbia were proven to be grounded in empirical realities, how, then, might we assume an interventionist stance that attempts the same "helpful advice" that Richard Florida and Charles Landry offer to spaces that aspire to be successfully (and perhaps also now anachronistically) urbane (Wilson and Keil 2009)? How, on the other hand, do we avoid the characteristic dangers of anti-urbanism—its unregulated sprawl, its disdain for institutions, its attempt to escape, its avoidance of challenges beyond the immediate?

Eschewing both these paths, extraurbanism needs to redefine and supplement the historic virtues of the urban, rather than renounce them. We need to develop a normativity for a genuinely transformational extraurbanism in which the urban is revived by collateral extraurban flows.

The case we have mounted in this chapter is conjectural. To attain cogency and to have purchase on our lived experience, it would need considerable empirical support, theoretical refinement and the development of an elaborated normative agenda. Some aspects of such a program would include the following:

1 *Recalculations of numbers in place.* With the space of the city redrawn, and with the urban biases removed from our categories and processes of quantification, just how many people on earth are, in our redefinition, urban dwellers? And are people still in fact heading to these urban spaces once we have reconfigured of our spatial categories? What, on the other hand, are the social and cultural demographics of extraurbia? How are they changing?
2 *Reassessment of qualities in space.* How does one measure differences in standards of living and qualities lifestyle between the urban and the extraurban? For instance, how does one calibrate lower costs of living against lower incomes? How does one figure immeasurable differences in qualities of dwelling? How does one read new patterns of self-realization, agency, and political engagement?
3 *Theorizing spaces and flows.* How do we define the extraurban? Who do we characterize the varieties of its spaces, the subtleties of its infinite variety? What, however, are the flows that constitute its symptomatic convergences? How do we test our conceptual conjectures?
4 *Framing scenarios.* What are the consequences of alternative extraurban trajectories and extraurban–urban dynamics? What happens in worst-case scenarios? What happens in strategically optimistic scenarios?

5 *Nurturing normativities.* It is one thing to read the changed shapes of spaces and directions of flows. It is another thing to know how to act in response. Do we go with the flows, speed the flows, slow the flows, or generate counterflows? These are questions for persons, households, policy makers, and social movements alike. Our normative interventions might range from "quiet encroachment" to vociferous engagement in the spaces of extraurbia.

References

Agyeman, J., and Neal, S. *The new countryside?: Ethnicity, nation and exclusion in contemporary rural Britain*. London: Policy Press.

Bayat, A. (2002). "Activism and social development in the Middle East." *International Journal of Middle East Studies*, 34, pp. 1–28.

Bell, D., and Jayne, M. (2009). "Small cities? Towards a research agenda." *International Journal of Urban and Regional Research*, 33, pp. 683–699.

Bernard, A. (2008). "Immigrants in the Hinterlands." *Perspectives on Labour and Income*, 9, pp. 5–14.

Bernard, R. M., Abrami, P. C., Lou, Y., Borokhovsk, E., Wade, A., Wozney, L., Wallet, P. A., Fiset, M., and Huang, B. (2004). "How does distance education compare with classroom instruction? A meta-analysis of the empirical literature." *Review of Educational Research*, 74, pp. 379–439.

Berry, Brian. (1976). "Urbanization and counter-urbanization." In *Urban Affairs Annual Reviews*. Thousand Oaks, CA: Sage.

Berube, A., Singer, A., Wilson, J. H., and Frey, W. H. (2006). *Finding exurbia: America's fast-growing communities at the metropolitan fringe*. Washington, DC: The Brookings Institution.

Brand, S. (2010). "How slums can save the planet." *Prospect Magazine*.

Brown, D. G., Johnson, K. M., Loveland, T. R., and Theobald, D. M. (2005). "Rural land-use trends in the conterminous United States, 1950–2000." *Ecological Applications*, 15, pp. 1851–1863.

Bruegmann, R. (2005). *Sprawl: A compact history*. Chicago: University of Chicago Press.

Cadieux, K. V., and Hurley, P. T. (2009). "Amenity migration, exurbia, and emerging rural landscapes: Global natural amenity as place and as process." *GeoJournal*, pp. 1–6.

Champion, A. G. (1992). *Counterurbanization: The changing pace and nature of population deconcentration*. London: Edward Arnold.

Champion, A. G., Hugo, G., and Champion, T. (2003). *New forms of urbanization: Beyond the urban-rural dichotomy*. New York: Ashgate.

Clancey, Gregory. (2004). "Local memory and worldly narrative: The remote city in America and Japan." *Urban Studies*, 41, pp. 2335–2355.

Collins, J. L., and Wellman, B. (2010). "Small town in the Internet society: Chapleau is no longer an island." *American Behavioral Scientist*, 53, pp. 1344–1366.

Cope, B., and Kalantzis, M. 2009. *Ubiquitous learning*. Urbana, IL: University of Illinois Press.

Cox, W. (2005). *Growth, economic development, and local government structure in Pennsylvania*. Belleville, IL: Wendell Cox Consultancy.

Cox, W. (2008). "America is more small town than we think." *New Geography, October 9.*

Davis, M. (2002). *Dead cities*. New York: The New Press.

Davis, M. (2006). *Planet of slums*. London: Verso.

Dear, M. J. (2000). *The postmodern urban condition*. Oxford, UK: Blackwell.

Ewing, R. H. (2008). "Characteristics, causes, and effects of sprawl: A literature review." *Urban Ecology*, pp. 519–535.

Fowler, G. A. (2010). "E-Commerce growth slows, but still out-paces retail." *Wall Street Journal*, March 8.

Garreau, J. (1991). *Edge city: Life in the new frontier*. New York: Doubleday.

Gimpel, J. G., and Lay, J. C. (2008). "Political socialization and reactions to immigration-related diversity in rural America." *Rural Sociology*, 73, pp. 180–204.

Gkartzios, M., and Scott, M. (2010). "Countering counter-urbanisation: Spatial planning challenges in a dispersed city-region, the Greater Dublin area." *Town Planning Review*, 81, pp. 23–52.

Glaeser, E. L., and Gyourko, J. (2002). "The impact of zoning on housing affordability." In *Policies to Promote Affordable Housing*, sponsored by the Federal Reserve Bank of New York and the New York University School of Law.

Glaeser, E. L., and Kohlhase, J. E. (2004). "Cities, regions and the decline of transport costs." *Papers in Regional Science*, 83, pp. 197–228.

Gosnell, H., and Abrams, J. (2009). "Amenity migration: Diverse conceptualizations of drivers, socioeconomic dimensions, and emerging challenges." *GeoJournal*, pp. 1–20.

Gunderson, R. J., Pinto, J. V., and Williams, R. H. (2008). "Economic or amenity driven migration? A cluster-based analysis of county migration in the Four Corners States." *Journal of Regional Analysis and Policy*, 38, pp. 243–254.

Hart, G. (2002). *Disabling globalization: Places of power in post-apartheid South Africa*. Berkeley, CA: University of California Press.

Harvey, D. (1996). *Justice, nature and the geography of difference*. Cambridge, MA: Blackwell.

Harvey, D. (2005). *A brief history of neoliberalism*. Oxford, UK: Oxford University Press.

Hunter, L. M. (2000). "A comparison of the environmental attitudes, concern, and behaviors of native-born and foreign-born US residents." *Population and Environment*, 21, pp. 565–580.

Jacobs, J. (1969). *The economy of cities*. New York: Random House.

Johnson, J. D., and Rasker, R. (1995). "The role of economic and quality of life values in rural business location." *Journal of Rural Studies*, 11, pp. 405–416.

Johnson, K. M., and Cromartie, J. B. (2006). "The rural rebound and its aftermath: Changing demographic dynamics and regional contrasts." Pp. 25–49 in *Population Change and Rural Society*, edited by W. A. Kandel and D. L. Brown. New York: Springer.

Jones, R. E., Fly, J. M., Talley, J., and Cordell, H. K. (2003). "Green migration into rural America: The new frontier of environmentalism?" *Society and Natural Resources*, 16, pp. 221–238.

Jones, S., and Fox, S. (2009). "Generations Online in 2009."

Kalantzis, M., and Cope, B. (2009). "Learner differences: Determining the terms of pedagogical engagement." Pp. 13–30 in *Beyond Pedagogies of Exclusion in Diverse Childhood Contexts*, edited by S. Mitakidou, E. Tressou, B. B. Swadener, and C. A. Grant. New York: Palgrave Macmillan.

Kotkin, J. (2009). "Hard times in the high desert: Exurbs may be struggling, but they're far from dead." *Forbes*, September 15.

Krannich, R. S., Petrzelka, P., and Brehm, J. M. (2006). "Social change and well-being in Western amenity-growth communities." Pp. 311–331 in *Population Change and Rural Society*, edited by W. A. Kandel and D. L. Brown. New York: Springer.

Krugman, P. (2005). "That hissing sound." *New York Times*, August 8.

Laliberte, N. (2009). "Sophistication at a country pace: Community sustainability and amenity-based development." *GeoJournal*, pp. 1–14.

Lang, R. (2003). *Edgeless cities: Exploring the elusive metropolis*. Washington, DC: Brookings Institution Press.

Larsen, S. C., Sorenson, C., McDermott, D., Long, J., and Post, C. (2007). "Place perception and social interaction on an exurban landscape in central Colorado." *The Professional Geographer*, 59, pp. 421–433.

Latour, B. (1996). *Aramis, or the love of technology*. Cambridge, MA: Harvard University Press.

Lawson, L. J. (2005). *City bountiful: A century of community gardening in America*. Berkeley, CA: University of California Press.

Lee, B. (2007_. "'Edge' or 'edgeless' cities? Urban spatial structure in U.S. metropolitan areas, 1980 to 2000." *Journal of Regional Science*, 47, pp. 479–515.

Lefebvre, H. 1970/2003. *The urban revolution*. Minneapolis: University of Minnesota Press.

Levitt, J. N. (2002). *Conservation in the Internet Age: Threats and opportunities*. Washington, DC: Island Press.

Lichter, D. T., Parisi, D., Grice, S. M., and Taquino, M. C. (2007). "National estimates of racial segregation in rural and small-town America." *Demography*, 44, pp. 563–581.

Mapes, J. E. (2009). *Urban revolution: Rethinking the American small town*. Los Angeles: University of Southern California.

Massey, D. (2005). *For space*. Los Angeles: Sage.

Miraftab, F., and Mcconnell, E. D. (2008). "Multiculturalizing rural towns: Insights for inclusive planning." *International Planning Studies*, 13, pp. 343–360.

Mitchell, W. J. (1999). *E-topia: Urban life, but not as we know it*. Cambridge, MA: MIT Press.

Mitchell, W. J., Borroni-Bird, C. E., and Burns, L. D. (2010). *Reinventing the automobile: Personal mobility in the 21st century*. Cambridge, MA: MIT Press.

Munkejord, M. C. (2006). "Challenging discourses on rurality: Women and men in-migrants' constructions of the good life in a rural town in northern Norway." *Sociologia Ruralis*, 46, pp. 241–257.

Nelson, L., and Hiemstra, N. (2008). "Latino immigrants and the renegotiation of place and belonging in small town America." *Social and Cultural Geography*, 9, pp. 319–342.

Ness, D. (2007). "Sustainable infrastructure: Doing more with less by applying eco-efficiency principles." Pp. 87–96 in *Sustainable Infrastructure in Asia*, Conference paper, United Nations Economic and Social Commission for Asia and the Pacific, September 6–8.

O'Toole, K., and Burdess, N. (2004). "New community governance in small rural towns: The Australian experience." *Journal of Rural Studies*, 20, pp. 433–443.

O'Toole, K., and Burdess, N. (2005). "Governance at community level: Small towns in rural Victoria." *Australian Journal of Political Science*, 40, pp. 239–254.

Peters, M. A., Marginson, S., and Murphy, P. (2008). *Creativity and the global knowledge economy*. New York: Peter Lang.

Pfeffer, M. J., and Stycos, J. M. (2002). "Immigrant environmental behaviors in New York City." *Social Science Quarterly*, 83, pp. 64–81.

Phelps, N. A., Wood, A. M., and Valler, D. C. (2010). "A postsuburban world? An outline of a research agenda." *Environment and Planning A*, 40, pp. 366–383.

Pieterse, J. N. (2008). "Globalization the next round: Sociological perspectives." *Futures*, 40, pp. 707–720.

Pink, S. (2009). "Urban social movements and small places: Slow cities as sites of activism." *City*, 13, pp. 451–465.

Pisarski, A. E. (2006). *Commuting in America*. Washington, DC: Transportation Research Board.

Quark, A. A. (2007). "From global cities to the lands' end: The relocation of corporate headquarters and the new company towns of rural America." *Qualitative Sociology*, 30, pp. 21–40.

Quiggin, J. (2010). *Zombie economics: How dead ideas still walk among us*. Princeton, NJ: Princeton University Press.

Qviström, M. (2007). "Landscapes out of order: Studying the inner urban fringe beyond the rural–urban divide." *Geografiska Annaler: Series B, Human Geography*, 89, pp. 269–282.

Rasker, R., Gude, P. H., Gude, J. A., and van den Noort, J. (2009). "The economic importance of air travel in high-amenity rural areas." *Journal of Rural Studies*, 25, pp. 343–353.

Ryker, L. (2007). *Off the grid homes: Case studies for sunstainable living*. Layton, UT: Gibbs Smith.

Sassen, S. (2006). *Territory – authority – rights: From medieval to global assemblages*. Princeton, NJ: Princeton University Press.

Schilling, J., and Logan, J. (2008). "Greening the rust belt: A green infrastructure model for right sizing America's shrinking cities." *Journal of the American Planning Association*, 74, pp. 451–466.

Schluter, G., and Lee, C. (2002). "Can rural employment benefit from changing labor skills in U.S. processed food trade?" *Rural America*, 17, p. 38–43.

Shachar, M., and Neumann, Y. (2003). "Differences between traditional and distance education academic performances: A meta-analytic approach." *The International Review of Research in Open and Distance Learning*, 4, pp. 1–20.

Shachar, M. and Neumann, Y. (2010). "Twenty years of research on the academic performance differences between traditional and distance learning: Summative meta-analysis and trend examination." *Journal of Online Learning and Teaching*, 6, pp. 318–334.

Short, J. R., Hanlon, B., and Vicino, T. J. (2007). "The decline of inner suburbs: The new suburban gothic in the United States." *Geography Compass*, 1, pp. 641–656.

Singer, A. (2004). *The rise of new immigrant gateways*. Washington, DC: Brookings Institution.

Soja, E. M. (2000). *Postmetropolis: Critical studies of cities and tegions*. Malden, MA: Blackwell.

Sööt, S., Berman, J. G., and DiJohn, J. (2006). "Emerging commuting trends: Evidence from the Chicago area." *Journal of the Transformation Research Forum*, 45, pp. 109–120.

Spectorsky, A. C. (1955). *The Exurbanites*. Berkeley, CA: Berkeley Publishing.

Storper, M. (1997). *The regional world: Territorial development in a global economy.* New York: Guilford.

Sullivan, S., McCann, E., and De Young, R. (1996). "Farmers' attitudes about farming and the environment: A survey of conventional and organic farmers." *Journal of Agricultural and Environmental Ethics*, 9, pp. 123–143.

Taylor, L. (2009). "No boundaries: Exurbia and the study of contemporary urban dispersion." *GeoJournal*, pp. 1–17.

Tongia, R., and Wilson, E. J. (2007). "Turning Metcalfe on his head: The multiple costs of network exclusion." In *Proceedings of the 35th Telecommunications Policy Research Conference.*

Vias, A. C., and Nelson, P. B. (2006). "Changing livelihoods in rural America." Pp. 75–102 in *Population Change and Rural Society*, edited by W. A. Kandel and D. L. Brown. New York: Springer.

Walker, P., and Fortmann, L. (2003). "Whose landscape? A political ecology of the 'exurban' Sierra." *Cultural Geographies*, 10, pp. 469–491.

Wilson, D., and Keil, R. (2009). "Mythologies and the creative city discourse." *Social and Cultural Geography*, 9(8), 841–847.

Woods, M. (2009). "The local politics of the global countryside: Boosterism, aspirational ruralism and the contested reconstitution of Queenstown, New Zealand." *GeoJournal*, pp. 1–17.

Acknowledgment

The authors wish to thank David Wilson and Fazal Rizvi for their inspiration and research assistant Erin DeMuynck for searching and sourcing.

Index

f indicates figure and *t* indicates table

Printed and bound by CPI Group (UK) Ltd, Croydon, CR0 4YY